Milestones in Drug Therapy

For further volumes:
http://www.springer.com/series/4991

Mark von Itzstein
Editor

Influenza Virus Sialidase - A Drug Discovery Target

Springer

Volume Editor
Prof. Mark von Itzstein
Director
Institute for Glycomics
Gold Coast Campus
Griffith University
QLD 4222
Australia
m.vonitzstein@griffith.edu.au

Series Editors
Prof. Michael J. Parnham, Ph.D.
Visiting Scientist
Research & Clinical Immunology Unit
University Hospital for Infectious Diseases "Dr. Fran Mihaljević"
Mirogojska 8
HR-10000 Zagreb
Croatia

Prof. Dr. Jacques Bruinvels
Sweelincklaan 75
NL-3723 JC Bilthoven
The Netherlands

ISBN 978-3-0348-0779-1 ISBN 978-3-7643-8927-7 (eBook)
DOI 10.1007/978-3-7643-8927-7

Printed on acid-free paper

Springer Basel AG is part of Springer Science + Business Media (www.springer.com)

Preface

Influenza virus remains a significant threat to humanity [1] despite the discovery of novel antiviral therapies and the continuing development of seasonal vaccines. Annual influenza epidemics exact a high toll in morbidity, estimated to be in the range of three to five million cases of severe illness, and mortality, with up to half a million deaths worldwide [2]. The twenty-first century's first influenza pandemic caused by the commonly referred 'swine flu' virus strain (H1N1) 2009 has had major economic impact across the world and accounted for 414,000 confirmed cases and 5,000 deaths worldwide [3]. Human infection with the highly pathogenic avian H5N1 influenza A virus accounted for 552 confirmed cases and 322 deaths worldwide between 2003 and 2011 [4]. A recent study has estimated that a human-to-human transmissible, highly pathogenic, pandemic influenza virus could lead to ~62 million deaths worldwide [5]. The prevalence of oseltamivir-resistant influenza A(H1N1) viruses increased to 99% in many countries during 2008–2009 flu season [6] and recent isolation of oseltamivir-resistant [7] H5N1 *thus necessitate the continued development of alternative antiviral agents.*

In this edited volume, Klenk provides an excellent review on the virology of influenza virus and its pathogenesis. As introduced in this particular contribution influenza virus sialidase (neuraminidase EC 3.2.1.18) is an enzyme (an *exo*-glycohydrolase) and is a tetrameric glycoprotein that consists of four identical subunits [8] anchored to the viral membrane by a long thin stalk. This enzyme effectively acts as a pair of biological scissors cleaving a-ketosidically linked sialic acids from glycoconjugates [8]. This enzyme action facilitates both the movement of virus particles through the upper respiratory tract, and, importantly, the escape of virion progeny from the surface of infected cells [8–11]. The essential nature of the influenza virus sialidase in the virus' life cycle makes it an ideal drug discovery target.

In subsequent contributions, Nicholls and colleagues provide an invaluable overview on the characterisation of carbohydrate recognition by influenza virus and a detailed analysis of virus tropism, Chan and Bennet elaborate on sialidase enzymology and Dyason and von Itzstein provide an up-to-date discussion on recent developments in influenza virus sialidase structure-based drug design.

Influenza virus sialidase [8] targeted anti-influenza drug discovery and development has used structure-assisted drug design to discover the first highly potent influenza virus sialidase inhibitor that mimics characteristics of the putative enzyme transition state: zanamivir (Relenza®) **1** [12, 13] and subsequently oseltamivir (Tamiflu®) **2** [14] are now marketed for treatment and prophylaxis of influenza virus infection. The worldwide stockpiling of these two antiviral drugs as part of pandemic preparedness highlights the overall importance of influenza virus sialidase inhibitors.

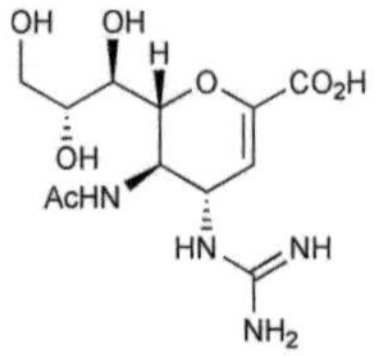

1
zanamivir

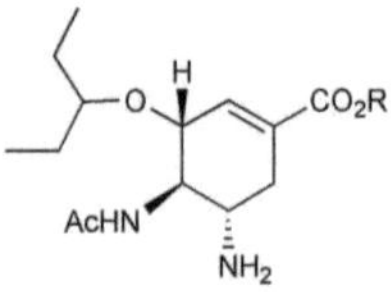

2 R = Et
oseltamivir

Substantial chemical methodologies have been developed in the discovery of both zanamivir [13] and oseltamivir [14]. The recent chemistries associated with carbohydrate-based and non-carbohydrate-based influenza virus sialidase inhibitor discovery are described, in good detail, by Thomson and von Itzstein and Streicher and Stanley, respectively. Both contributions highlight the latest approaches towards the synthesis of influenza virus sialidase inhibitors.

In the clinical setting, both zanamivir and oseltamivir are effective in both the prevention and treatment of influenza A and B infection. Benefit in treatment is typically restricted to patients treated within 48 h of symptom onset [15]. The influenza virus surface glycoproteins haemagglutinin and sialidase continually undergo natural and antibody-induced antigenic drift [16]. However, as the active site residues of influenza virus sialidase that are involved in interactions with the designed sialidase inhibitors are highly conserved, it was hoped the likelihood of variants with uncompromised infectivity and transmissibility would be reduced. The final two contributions by Buchy and Naffakh and colleagues provide excellent insight into the clinical experience of influenza virus sialidase targeting anti-influenza drugs and the complications of resistance development to these drugs respectively.

In summary, in the first part of this edited volume readers are provided with an excellent and comprehensive multi-disciplinary overview of (1) influenza virus sialidase as an essential enzyme in the lifecycle of the virus, (2) the importance of carbohydrate recognition phenomena for successful virus infection of host cells, (3) influenza virus sialidase enzymology and (4) influenza virus sialidase structure-based drug discovery. Furthermore, in the second part the reader is provided with a substantial review of the chemistry associated with influenza virus sialidase inhibitor development, the importance of the lead anti-influenza drugs zanamivir and oseltamivir in a clinical setting and the potential of the virus to develop resistance to the first-line-of-defence anti-influenza drugs.

Gold Coast, Queensland,
Australia

Mark von Itzstein

References

1. Zambon M, Hayden FG, Global Neuraminidase Inhibitor Susceptibility Network (2001) Position Statement: global neuraminidase inhibitor susceptibility network. Antiviral Res 49(3):147–156
2. Fiore AE, Shay DK, Haber P, Iskander JK, Uyeki TM, Mootrey G, Bresee JS, Cox NJ, Advisory Committee on Immunization Practices (ACIP), Centers for Disease Control and Prevention (CDC) (2007) Prevention and control of influenza. Recommendations of the Advisory Committee on Immunization Practices (ACIP) MMWR Recommend. Rep 56(RR-6):1–54
3. WHO pandemic (H1N1) 2009-update 71
4. Confirmed Human Cases of Avian Influenza A(H5N1) http://www.who.int/csr/disease/avian_ influenza/ country/en.
5. Murray CJ, Lopez AD, Chin B, Feehan D, Hill KH (2006) Estimation of potential global pandemic influenza mortality on the basis of vital registry data from the 1918–20 pandemic: a quantitative analysis. Lancet 368:2211–2218
6. Sheu TG, Fry AM, Garten RJ, Deyde VM, Shwe T, Bullion L, Peebles PJ, Li Y, Klimov AI, Gubareva LV (2011) Editor's choice: dual resistance to adamantanes and oseltamivir among seasonal influenza A(H1N1) viruses: 2008–2010. J Infect Dis 203(1):13–17
7. Le QM, Kiso M, Someya K, Sakai YT, Ngyyen TH, Nguyen KH, Pham ND, Ngyen HH, Yamada Sm Muramoto Y, Horimoto T, Takada A, Goto H, Suzuki T, Suzuki Y, Kawaoka Y (2005) Avian Flu: isolation of drug-resistant H5N1 virus. Nature 437:1108
8. Colman PM, Ward CW (1985) Structure and diversity of influenza virus neuraminidase. Curr Top Microbiol Immunol 114:177–255
9. Palese P, Tobita K, Ueda M, Compans RW (1974) Characterization of temperature sensitive influenza virus mutants defective in neuraminidase. Virology 61:397–410
10. Liu C, Eichelberger MC, Compans RW, Air GM (1995) Influenza type A virus neuraminidase does not play a role in viral entry, replication, assembly, or budding. J Virol 69:1099–1106
11. Herrler G, Hausmann J and Klenk H-D (1995) Sialic acid as receptor determinant of Ortho- and Paramyxoviruses. In Rosenberg A (ed) Biology of the sialic acids. Plenum Press, New York, pp 315–336
12. von Itzstein M (2007) The war against influenza: discovery and development of sialidase inhibitors. Nat Rev Drug Discov 6:967–974
13. von Itzstein M, Wu WY, Kok GB, Pegg MS, Dyason JC, Jin B, Van Phan T, Smythe ML, White HF, Oliver SW, Colman PM, Varghese JN, Ryan DM, Woods JM, Bethell RC, Hotham VJ, Cameron JM, Penn CR (1993) Rational design of potent sialidase-based inhibitors of influenza virus replication. Nature 363:418–423

14. Kim CU, Lew W, Williams MA, Liu H, Zhang L, Swaminathan Sm, Bischofberger N, Chen MS, Mendel DB, Tai CY, Laver WG, Stevens RC (1997) Influenza neuraminidase inhibitors possessing a novel hydrophobic interaction in the enzymen active site: design, synthesis and structural analysis of carbocyclic sialic acid analogues with potent anti0influenza activity. J Am Chem Soc 119:681–690
15. Fleming DM (2003) Zanamivir in the treatment of influenza. Expert Opin Pharmacother 4:799–805
16. Webster RG, Bean WJ, Gorman OT, Chambers TM, Kawaoka Y (1992) Evolution and ecology of influenza A viruses. Microbiol Rev 56:152–179

Contents

Influenza Virology

Hans Dieter Klenk

Abstract Influenza virus, comprising types A, B, and C, belong to the *Orthomyxoviridae* family. Influenza A viruses are important human and animal pathogens. Their natural hosts are aquatic birds from which they are occasionally transmitted to other species. In man, they cause outbreaks of respiratory disease that occur as annual epidemics and less frequent pandemics. Virus particles are composed of a lipid envelope with glycoprotein spikes and a nucleoprotein core containing the negative-stranded RNA genome and the RNA-dependent RNA polymerase. Reassortment of the segmented genome and a high mutation rate account for the unusual variability of influenza A viruses. The 8 RNA segments encode 11 proteins. The haemagglutinin (HA) is responsible for virus entry into the host cell by binding to sialic acid-containing receptors and by inducing membrane fusion, the viral polymerase mediates RNA replication and transcription in the nucleus, and the neuraminidase promotes release of progeny virions that are formed by budding from the cell surface. The non-structural NS1 protein has several functions, including that of an interferon antagonist. Host specificity and pathogenicity result from the interaction of numerous host factors with all viral proteins among which, however, the polymerase, the HA, and NS1 play particularly prominent roles.

1 Introduction

Influenza viruses were discovered in 1931 in pigs [1] and shortly thereafter in man [2]. At that time, fowl plague virus (FPV) had already been known for 20 years as one of the first viral agents ever isolated [3, 4]), but it took another 20 years until it was identified as an avian influenza virus [5]. Thus, the discovery of influenza viruses suggested already a zoonotic scenario, and it is now widely accepted that

H.D. Klenk (✉)
Institut für Virologie, Hans-Meerwein-Str. 2, 35043, Marburg, Germany
e-mail: klenk@staff.uni-marburg.de

M. von Itzstein (ed.), *Influenza Virus Sialidase - A Drug Discovery Target*,
Milestones in Drug Therapy, DOI 10.1007/978-3-7643-8927-7_1,

human influenza viruses emerge from a large reservoir of animal viruses (for review, see [6]).

Early studies on influenza viruses were also instrumental for the development of the general concept that interaction with specific host cell receptors plays an important role in the life cycle of a virus. Hirst [7] and McClelland and Hare [8] discovered that influenza virus is able to haemagglutinate and that adsorbed virus is eluted from erythrocytes on incubation at 37°C indicating an enzymatic destruction of a receptor substance on the cells. The enzymatic activity of influenza virus resembled the receptor destroying enzyme (RDE) of *V. cholerae* [9] which was therefore also used as a tool in the search for the receptor. The substance released by the viral enzyme was initially characterized as a carbohydrate of low molecular weight [10] and then identified in crystalline form as *N*-acetyl-D-neuraminic acid [11]. Thus, it was clear that the receptor determinant of influenza virus was sialic acid and that the viral enzyme was a neuraminidase (sialidase, NA). As will be illustrated in this and the following chapters of this book it is now well established that NA, like the haemagglutinin (HA), is a crucial determinant for host range and pathogenicity of influenza viruses as well as an important target for therapeutic intervention.

2 Classification

The influenza viruses are members of the *Orthomyxoviridae* family. Based on antigenic differences of their nucleocapsid and matrix proteins, three different types (genera) can be distinguished: influenza A, B, and C viruses. Influenza A viruses are further divided into subtypes characterized by 16 different haemagglutinins (H1–H16) and nine different neuraminidases (N1–N9). The designation of individual strains indicates type, host of origin (except for human isolates), site of isolation, strain reference number, year of isolation, and HA and NA subtypes: A/Hamburg/6/06 (H3N2) and A/duck/Minnesota/1525/81 (H5N1) are therefore examples of a human and an avian isolate, respectively.

3 Virus Structure and Genome Organization

Influenza viruses are pleomorphic, enveloped particles (Fig. 1). Spherical influenza virus particles have a diameter of approximately 100 nm, but long filamentous particles (from 300 nm to several micrometers in length) have also been observed, particularly in fresh clinical isolates. The viral glycoproteins are embedded in the host-derived lipid envelope and are visible as spikes that radiate from the exterior surface of the virus when particles are viewed under the electron microscope. For influenza A and B viruses, the glycoproteins consist of the HA protein, which is the major surface protein and the NA protein. Influenza C viruses have only one surface glycoprotein, the HEF (haemagglutinin/esterase/fusion) protein. Additional minor

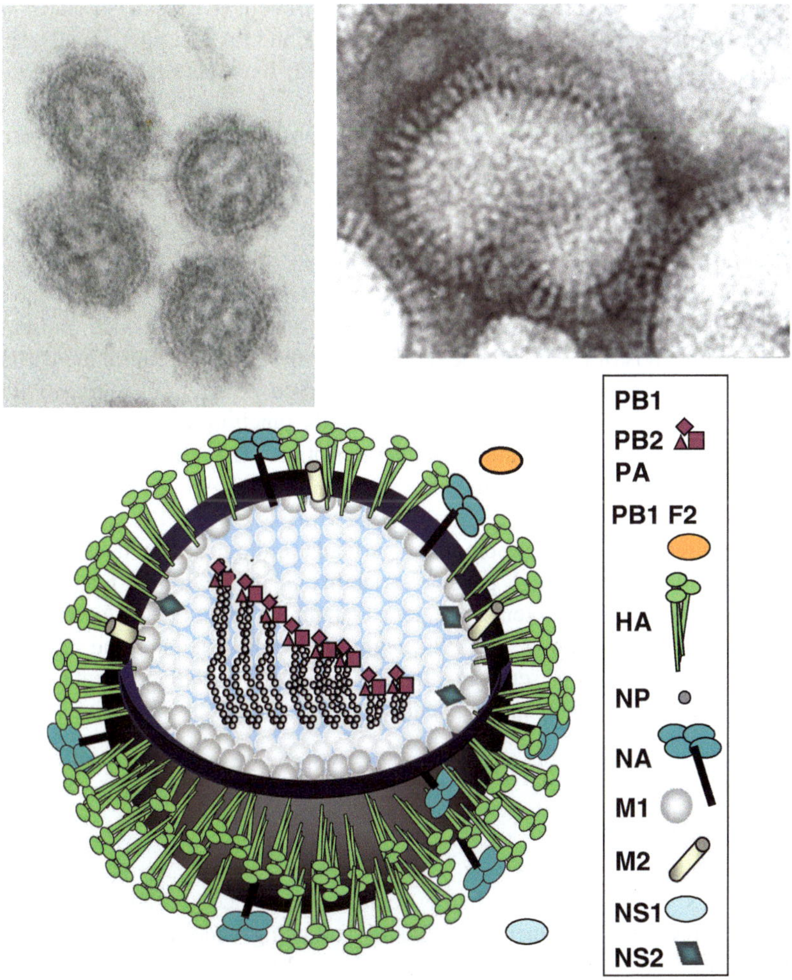

Fig. 1 Structure of influenza A virus. The electron micrograph on the left side shows thin sections of fowl plague virus particles. Virions have a diameter of 100 nm. Inside particles, cross sections of nucleocapsid segments can be seen. The electron micrograph on the right side shows a virus particle visualized by negative staining. The glycoprotein spikes (10 nm in length) protruding at the surface can be seen. Virus model: The envelope contains haemagglutinin (HA) and neuraminidase (NA) spikes as well the M2 protein. The interior of the particle is formed by ribonucleoproteins, consisting of the genomic RNA segments, the nucleocapsid protein (NP), and the polymerase proteins PB1, PB2, and PA. The inner side of the envelope is lined by the matrix protein M1. Virions contain also small amounts of the NS2 protein. PB1-F2 and NS1 are non-structural proteins

components of the viral envelope are the M2 protein (influenza A viruses), NB and BM2 proteins (influenza B viruses), and the CM2 protein (influenza C viruses). The most abundant virion protein M1 makes up the matrix which lies beneath the lipid membrane and surrounds the ribonucleoprotein (RNP) complexes. RNPs consist of the viral RNAs which are coated with the nucleoprotein (NP) and associated with the heterotrimeric polymerase complex (PB1, PB2, and PA proteins). Finally, small amounts of the nuclear export protein (NEP/NS2) have also been found within influenza A and B virus particles [12].

The genomes of influenza A and B viruses consist of eight separate negative-sense, single-stranded RNA segments known as vRNA. The influenza C virus genome has seven such segments. The genomes of influenza A, B, and C viruses have a total length of about 13,600, 14,600, and 12,900 nucleotides, respectively. Each viral RNA segment exists as an RNP complex in which the RNA is coated with nucleocapsid protein (NP) and forms a helical hairpin that is bound on one end by the heterotrimeric polymerase complex. Non-coding sequences are present at both 5′ and 3′ ends of the viral gene segments. At the extreme termini are partially complementary sequences that are highly conserved between all segments in all influenza viruses. When base paired, these ends function as the viral promoter which is required for replication and transcription. The additional non-coding sequences contain the polyadenylation signal for mRNA synthesis as well as parts of the packaging signals required during virus assembly [12].

The segmentation of the genome is of particular biological significance, since it is a precondition for the high genetic variability of these viruses. Each segment encodes one or two viral proteins. The gene assignment for influenza A viruses is as follows: RNA segment 1 codes for PB2, segment 2 for PB1 and in some strains also PB1-F2, segment 3 for PA, segment 4 for HA, segment 5 for NP, segment 6 for NA, segment 7 for M1 and M2, and segment 8 for NS1 and NS2/NEP [12].

4 Life Cycle

HA initiates infection by mediating binding to *N*-acetyl-D-neuraminic acid-containing receptors and membrane fusion following endocytosis. The M2 ion channel, associated with influenza A viruses only and not B virus, plays an important role in uncoating by lowering the pH within the virus particle and, thus, allowing dissociation of the internal components. The uncoating can be inhibited by amantadine and rimantadine which are approved anti-influenza drugs targeting the M2 protein. These compounds are only effective against influenza A viruses. The nucleocapsid complexes are then transported into the nucleus, where transcription and replication take place. The genomic RNA (vRNA) serves as template for two different RNA species: complementary RNA (cRNA) which is a complete copy of vRNA, and mRNA with a cap structure at the 5′ end and with the 3′ terminal nucleotides of cRNA being replaced by a poly-A-tail. The cRNA is the template for new vRNA molecules. The cap structures of the viral mRNAs are derived from cellular mRNA molecules. Viral mRNA utilizes the cellular translation machinery for the synthesis of viral proteins. The virus enhances the coding capacity of its genome by several mechanisms. These include splicing, expression of two cistrons in tandem position, and start of translation at two different initiation codons. Translation of NP, polymerase proteins, matrix protein, and proteins NS1 and NS2 occurs on free polysomes. Ribonucleoproteins are assembled in the nucleus and subsequently exported into the cytoplasm. The envelope proteins are translated at the rough endoplasmic reticulum and then transported by the exocytotic apparatus to the plasma membrane where virus particles are assembled in a budding process [12]. NA mediates virus release by removing receptors from the infected cell, but it plays also a role in virus entry (for references see below). The NA inhibitors zanamivir and oseltamivir are approved drugs that have strong prophylactic and therapeutic activity against influenza A and B viruses.

5 Ecology and Evolution

The high genetic variability of influenza viruses is the prime determinant for the complex ecology and epidemiology of these agents. This is most obvious with the influenza A viruses with their large number of HA and NA subtypes. Whereas only some of these subtypes have been observed with man, swine, horse, and other mammals, the entire spectrum is found in birds. In fact, it is now widely accepted that aquatic birds of the orders *Anseriformes* (waterfowl) and *Charadriformes* (shore birds) are the natural hosts of the influenza A viruses [6]. Of the 144 possible combinations of subtypes, 105 have been detected in these animals [13].

Most avian viruses cause no or only mild disease symptoms in birds. These low pathogenic avian influenza (LPAI) viruses comprising all subtypes have to be discriminated from the highly pathogenic avian influenza (HPAI) viruses. HPAI viruses are defined as H5 and H7 viruses that cause 75% or higher mortality after experimental infection of chickens and have a polybasic HA cleavage site (see below). All viruses that do not meet these criteria are LPAI viruses [14].

HPAI viruses arise by introduction and circulation of H5 and H7 LPAI viruses in domestic poultry with subsequent mutations in HA [15, 16]. In general, aquatic birds are not the natural host of HPAI viruses. However, there is evidence that the H5N1 HPAI viruses circulating in Asia have recently been reintroduced from poultry into wild aquatic birds [17].

Since the host barrier is not an insurmountable obstacle for LPAI viruses, they can occasionally be transmitted from their natural reservoir to terrestrial birds and mammals. Most of these transmissions are transient and do not result in a stable new lineage. On rare occasions, however, the viruses adapt to the new species and thus give rise to a new lineage. Adaptation requires multiple mutations and it may involve gene reassortment after co-infection with another virus.

There is evidence that viruses originating from aquatic birds acquire specificity for human receptors (see below) when replicating in chickens and quails as intermediate hosts [18, 19]. These viruses may then infect humans and undergo further adaptation with mutations in genes that encode both surface and internal proteins, or reassort with a human virus acquiring genes that encode internal proteins required for efficient viral replication [20]. In another scenario, avian viruses are transmitted to pigs, where they adapt for efficient growth in humans by acquiring again mutations in HA that are necessary for recognition of human receptors and possibly also in the genes encoding internal proteins [21]. Alternatively, pigs serve as a mixing vessel in which both avian and human viruses reassort to give rise to a new human virus [22, 23]. By these mechanisms, viruses with new surface glycoproteins and therefore a distinct change in antigenicity may be generated. If a new virus with such an antigenic shift occurs in man, it causes a pandemic.

6 Epidemiology

Human influenza viruses are responsible for periodic seasonal epidemics and occasionally for pandemics that are less frequent, but, in general, more severe. Epidemics are caused by influenza A and B viruses, whereas pandemics are only caused by influenza A viruses.

6.1 Pandemics

It has been estimated that there have been at least 13 pandemics in the last 500 years [24] including five scientifically well documented ones since 1918 (Fig. 2).

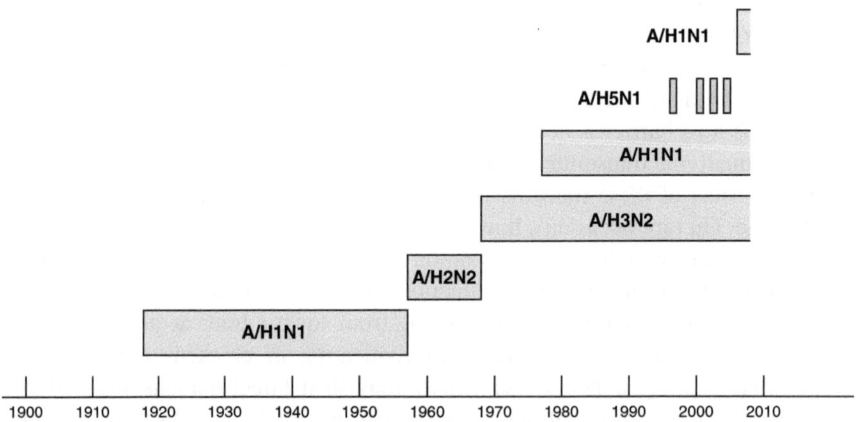

Fig. 2 Human influenza A viruses between 1918 and 2009. H1N1, H2N2, and H3N2 periods are shown. Sporadic H5N1 transmissions to humans are also indicated

The Spanish influenza (H1N1). This pandemic was the most devastating one taking the lives of about 50 million people world wide. It emerged in the spring of 1918 and swept within 1 year around the globe in three waves. Whereas the first wave was relatively mild, severity increased in the two subsequent ones with mortality rates of about 3%, as compared to less than 0.1% typical for ordinary outbreaks. With a high mortality rate among young adults the age-dependent mortality pattern was also unusual. The major cause of death was pneumonia that was often complicated by bacterial superinfection. Reconstruction of the virus from archival patient samples revealed that it was presumably derived without reassortment from an avian H1N1 virus [25–27] and allowed studies that threw light on some of its pathogenetic properties. It was found that HA [28–30], the replication complex [27, 31, 32], NS1 [33], and PB1-F2 [34] contributed to virulence, and that HA and PB2 determined transmissibility [31]. It was also found that the virus elicits aberrant immune response in animals [35, 36] which may also contribute to its high pathogenicity in man.

The Asian influenza (H2N2). This pandemic emerged in Southern China early in 1957 and spread over other parts of East Asia to North America and Europe where it appeared in the fall. The virus was derived from the previous H1N1 virus that had acquired HA and NA genes of subtypes H2 and N2, respectively, as well as its PB1 gene from avian viruses by reassortment [37, 38]. With about 1 million deaths worldwide the H2N2 virus was significantly less pathogenic than the 1918 virus.

The Hong Kong influenza (H3N2). This virus was first isolated in July 1968 in Hong Kong, and the pandemic peaked in the following two winters. The H3N2 virus again was the result of gene reassortment between the preceding H2N2 virus and avian viruses providing the HA and PB1 genes [37, 38]. In terms of mortality, the Hong Kong pandemic, like the previous one, was relatively mild.

The Russian influenza (H1N1). In May 1977, H1N1 virus reappeared in an outbreak that originated in the Russian-Chinese border region and reached North America in 1978. The virus which closely resembled isolates obtained in the 1950s [39, 40] did not replace the H3N2 in 1977, and both subtypes are co-circulating ever since.

The outbreak of 1946–1947 (H1N1). Looking at the outbreaks of 1918, 1957, 1968, and 1977 one may conclude that pandemics occur when viruses with a new HA subtype, or new HA and NA subtypes, emerge in the human population. It cannot be excluded, however, that a pandemic virus is also generated when it acquires by reassortment a novel NA alone or an HA that is of the same serotype as the one in the circulating virus but antigenically remote from it [41]. In fact, the influenza outbreak of 1946–1947 was originally considered a pandemic, because there was no serological cross reaction between the new agent and its predecessors [42]. Moreover, vaccination against these viruses did not confer protection against infection by the new strain [43]. The 1946–1947 virus lost its pandemic status when it became clear that it belonged to the H1N1 subtype as did the viruses in the preceding years [44]. However, a more recent study employing extensive antigenic characterization combined with sequence analysis of the glycoproteins of viruses obtained in 1943 and 1947 revealed marked antigenic and structural differences particularly in the HAs of these viruses [45]. These observations indicated that the 1947 virus was not an antigenic drift variant (see below) of the 1943 virus. They rather suggest that at that time the seasonal strains of the preceding years were replaced by another, yet only distantly related, H1N1 virus that was either generated by intrasubtypic reassortment or by introduction in toto from an unknown source.

The swine-origin 2009 influenza (H1N1). The concept that a pandemic may not necessarily depend on the introduction of a virus with an HA subtype that is new to the human population is further supported by the influenza outbreak in 2009 caused by an H1N1 virus presumably of swine origin. The virus emerged in February 2009 in Mexico and has swept within the next 6 months around the globe. In general, the disease has been mild, but there have also been cases requiring intensive care treatment that included the use of NA inhibitors. Whether the outbreak increases in severity and how it will compare to previous pandemics remains to be seen. The virus probably resulted from the reassortment of recent North American H3N2 and H1N1 swine viruses with Eurasian avian-like swine viruses. Its PB2 and PA genes originate from North American avian viruses; the PB1 gene from a human H3N2 virus; the HA, NP, and NS genes from classical swine viruses: and the NA and M genes from Eurasian avian-like swine viruses [46]. The virus has retained some genetic host range markers typical for avian viruses suggesting that it is still able to further adapt to humans and that it has not exhausted yet its pathogenetic potential.

6.2 Seasonal Epidemics

Seasonal epidemics are usually milder than pandemics, but still have a high impact on public health. The viruses responsible for epidemics are derived from pandemic viruses by antigenic drift. Antigenic drift is defined as gradual minor antigenic changes that are caused by point mutations in the HA or NA protein. Drift variants spontaneously selected in the presence of neutralizing antibodies are therefore antibody escape mutants. Mutations in HA or NA of human influenza viruses occur at a frequency of less than 1% per year. Nonetheless, antigenic drift variants can cause epidemics and often prevail for 2–5 years before being replaced by a different variant.

HA is the major antigenic spike glycoprotein. Elucidation of the atomic structure of HA and mapping of both natural HA variants and laboratory-derived monoclonal antibody escape mutants revealed that HA has five antigenic sites epitopes. All antigenic sites are located in the globular head of HA1. Antigenic drift has also been found in NA, and the sites of antigenic drift mapped to specific regions of the NA atomic structure.

6.3 Transmission of Avian Influenza Viruses to Man

For a long time, HPAI viruses have been thought to be apathogenic for man. However, within the past years human infections with H5 and H7 HPAI as well as H9N2 LPAI viruses have been observed that were often associated with severe disease. As yet, none of them has adapted to man. However, since humans have not been exposed before and are therefore immunologically naïve to these viruses, they are a considerable pandemic threat.

Transmission of H5N1 HPAI virus was first observed during an epizootic in 1997 in Hong Kong, when 18 people became ill showing the symptoms of classical influenza. Six of these patients died. The virus that was transmitted without reassortment from infected chickens contained an HA gene derived from a H5N1 virus first isolated from a goose in 1996 (A/goose/Guangdong/1/96) [47–49]. Related H5N1 viruses circulated in birds in Southern China from 1997 to 2001 without transmission to humans [50, 51]. Since 2003 severe outbreaks have been observed periodically in Vietnam, Thailand, Indonesia, China, and other Asian countries [52]. Genetic comparison with the original isolates from 1997 revealed that the new viruses represented different genotypes resulting from multiple reassortments. Among these, the Z genotype was dominant. The HA of all of these viruses was similar to that of the A/goose/Guangdong/1/96 strain [53]. Extensive molecular–epidemiological studies revealed that domestic ducks and land-based poultry in Southern China probably had a central role in the generation and maintenance of the H5N1 virus and that wild birds might have contributed to the broad spread of the virus [20]. The role of migratory birds as carriers of the virus was further underlined when, following an outbreak in a bird sanctuary at Lake Qinghai in Northwest China in spring 2005 [54], the virus spread within a few months to Siberia, Europe, and Africa. It has to be pointed out, however, that international poultry trade may have played an important role in virus dissemination as well. As in 1997, the current H5N1 viruses have the capacity to cross species barriers and to cause disease with high case-fatality rates in mammals, such as cats [55], tigers [56], stone martens [57], and man. Human infection is characterized by severe pneumonia with high viral load and hypercytokinemia. There is also evidence for systemic virus spread [58].

Although sporadic human transmissions of HPAI H7N7 have been reported before [59, 60], there has been only one poultry outbreak in 2003 in the Netherlands where a larger number of human infections have been observed. Eighty-three cases with relatively mild influenza-like illness and conjunctivitis were observed. However, there was one human infection with acute respiratory distress syndrome and fatal outcome [61–63]. Seroepidemiological studies suggested that the number of inapparent infections was much larger, and there was also evidence for human-to-human transmission [64].

Human H9N2 infections with mild respiratory disease were observed in small numbers in 1999 and 2003 [52, 65, 66]. One of these viruses appears to be related to the HPAI H5N1 virus from 1997, since both have internal proteins with the same genetic background [67]. The H9N2 viruses are now enzootic in Eurasia [68].

7 Pathogenesis

7.1 Human Disease

Human influenza is a highly contagious disease that is transmitted by aerosols. Infection is usually confined to the respiratory tract. Replication occurs in the epithelia of the respiratory tract and reaches its peak 2–3 days after infection. The period of virus shedding is 7 days and may last with infants and children as long as 2 weeks. After an incubation time of 1–5 days, patients suffer from acute respiratory disease symptoms with headache, high fever, myalgia, nausea, and malaise. Severe cases develop typically primary influenza pneumonia or combined viral–bacterial pneumonia. Persons at risk for complications are the elderly and patients with cardiovascular diseases, metabolic disorders, or immunosuppression. Less frequent complications are myositis, myocarditis, and Reye's syndrome, a severe disease with brain and liver involvement and usually lethal outcome.

7.2 Avian Disease

LPAI virus replication is confined to the intestinal and respiratory tract resulting in mild disease or asymptomatic infection. Virus is shed in the feces. A common route of dissemination among aquatic birds is therefore through contaminated water. HPAI viruses are also shed in faeces with high concentrations. However, these viruses are more readily transmitted among birds in densely populated flocks by the nasal and oral routes through contact with virus-contaminated materials. In contrast to the local LPAI virus infection of the intestinal or respiratory tract, HPAI viruses cause systemic infection. As a result, virus can be recovered from many organs of infected animals. Large haemorrhages distributed all over the body, oedema, and cutaneous ischemia are major symptoms of the disease. The final stage of the infection can be characterized by the emergence of neurological signs, such as photobia and dullness [69, 70]. HPAI viruses have been found to specifically target lymphocytes and lymphoid tissues [71, 72], myocytes in the heart muscle [73], and endothelial cells [74], and these cell tropisms may play an important pathogenic role in systemic virus dissemination and in the vascular leakage underlying haemorrhages and oedema.

8 Role of Viral Proteins in Replication, Host Adaptation, and Pathogenesis

Influenza virus replication depends on the biosynthetic machinery of the cell and it is under the control of the defence mechanisms developed by the host against infection. Host specificity and pathogenicity are therefore the result of the interaction of numerous host factors with all viral proteins among which, however, the polymerase complex, HA and NS1, appears to play particularly prominent roles.

8.1 The Polymerase

The polymerase is a heterotrimeric complex consisting of three subunits, the PB2, PB1, and PA proteins [75]. The polymerase is active in the nucleus, and nuclear localization sequences have been identified on each subunit. Recent results suggest that cytoplasmically expressed PB1 and PA may be imported as a sub-complex, which then assembles with separately imported PB2 [76]. Several nuclear and cytoplasmic host proteins have been shown to be cofactors required by the polymerase [77–79].

The viral polymerase catalyses both the transcription of genomic RNA, i.e. viral RNA (vRNA), to mRNA and the replication of vRNA to complementary RNA (cRNA), the intermediate template which is necessary for copying new vRNA molecules [80]. Transcription is initiated by binding cap structures at the 5′ ends of cellular mRNAs by PB2 [81]. Capped 5’ mRNA ends are then cut off by an endonuclease activity of PB1 [82] and used as primers for further elongation which is eventually terminated due to stuttering [83, 84] at the uridine stretch adjacent to the 5′ end of the vRNA [85] leading to polyadenylation. In contrast, the replication of vRNA is primer independent. It involves the de novo initiation from nascent RNA chains and requires first the synthesis of cRNA, which then serves as template for new vRNA [80, 86].

As an RNA-dependent RNA polymerase the influenza virus polymerase lacks any proof-reading function. Thus, point mutations are introduced at a high rate, a prerequisite for high genomic plasticity upon selective pressure. Beside point mutations, major genetic changes such as deletions or insertions were reported with circulating viruses in the field. These include deletions within the stalk region of the NA [87–91] and insertions into the HA [92, 93]. Moreover, there are also some examples of intragenic recombination with other influenza gene segments [94–97]. However, insertion, deletion, and recombination in influenza A viruses are rare as compared to genetic changes due to point mutations and reassortment.

Pathogenicity studies in mice and in ferrets revealed several polymerase mutations responsible for increased virulence and enhanced replication. A single amino acid substitution in the PB2 protein, E-K at position 627, enables an avian virus to replicate in mammalian cells under experimental conditions. Furthermore, whereas in every avian influenza A virus PB2 protein the amino acid at position 627 is glutamic acid, in human strains it is always lysine [98]. This mutation increases the polymerase activity in mammalian cells. Moreover, polymerase complexes derived from avian viruses exhibited cold sensitivity in mammalian cells mostly due to the residue 627 in PB2 [99]. The amino acid substitution PB2 E627K was also found to contribute to enhanced virulence of several highly pathogenic avian influenza viruses of the H5N1 subtype from Southeast Asia both in mice [100–103] and in ferrets [104]. Several but not all H5N1 viruses isolated from humans carry this substitution [58, 105]. The relevance of this mutation in the field was underlined by its occurrence in another unrelated group of HPAIV of the H7N7 subtype. This mutation was found in an isolate from a deceased veterinarian during the fowl plague outbreak in the Netherlands in 2003 [61]. The characterization of the reconstructed PB2 sequence of the 1918 virus revealed the presence of 627K as well [106]. Moreover, several studies demonstrated the substitution PB2 D701N to increase the virulence of H7N7 and H5N1 HPAIV in mice [21, 107]. Other residues correlating with high pathogenicity are PB2 714R and PA 615N [21]. All these virulence enhancing mutations found in H7N7 and H5N1 viruses were shown by minireplicon assays to increase the polymerase activity in mammalian cells [102, 108, 109].

There is growing evidence that the increase in polymerase activity is accomplished by improved interaction of the polymerase proteins with host factors. Thus, it has been shown in a mouse model that the adaptive mutation D701N facilitates binding of PB2 to importin α, a component of the nuclear pore complex, thereby facilitating transport of this polymerase subunit into the nucleus where replication and transcription take place. Adaptation of the viral polymerase to the nuclear transport machinery plays therefore an important role in interspecies transmission of influenza virus [110]. Elucidation of the polymerase structure [111, 112] and identification of other host factors interacting with the polymerase [113] will throw more light on these adaptation mechanisms.

8.2 PB1–F2

The PB1 gene of most avian and human influenza A viruses encodes a second protein, PB1–F2, that is expressed from another reading frame. PB1-F2 is found in mitochondria, induces apoptosis [114], enhances inflammation in mice, and increases the frequency and severity of secondary bacterial infections [34].

8.3 The Nucleocapsid Protein (NP)

The main function of the highly basic protein NP is encapsidation of the viral RNA (NP monomer binds approximately 24 nt of RNA) which is necessary for recognition by the polymerase. NP also plays a crucial role in transporting the viral RNPs into the nucleus which it achieves through interaction with the host nuclear import machinery [12].

8.4 Haemagglutinin

The major surface glycoprotein HA is integrated in the virus envelope as a type I membrane protein. More than 7,500 nucleotide sequences of HA of various virus isolates comprising serotypes H1 to H16 have been determined, among them more than 900 HA sequences of H5N1 virus isolates alone [115]. The three-dimensional structure of several serotypes has been analysed by X-ray crystallography: H1 [116, 117], H3 [118, 119], H5 [120, 121], H7 [116], and H9 [120]. The ectodomain of HA represents 90% of the polypeptide chain. The residual 10% of the HA sequence account for the transmembrane domain and the cytosolic domain. HA is synthesized as a precursor molecule HA0 (75 kDa) which assembles to homotrimers. HA0 is N-glycosylated, palmitoylated, and proteolytically cleaved by host enzymes. Crystallographic analysis of the uncleaved precursor has shown that the cleavage site is located in a loop formed by 19 amino acids of which 8 amino acids protrude from the surface of the membrane-proximal third of the HA trimer [118]. There are differences in the orientation of the loop with serotypes H1 and H3 [117] both of which contain a monobasic cleavage site. Structural data are not available for the loop of HPAI viruses, but it can be assumed that it is extended by the insertion of the multibasic cleavage site (see below). The amino-terminal cleavage fragment HA1 (50kDa) contains the receptor binding site, and the carboxy-terminal fragment HA2 (25kDa) is membrane anchored and responsible for fusion.

8.4.1 Receptor Binding

The receptor determinant of influenza viruses is sialic acid (Sia), mostly *N*-acetyl-D-neuraminic acid (Neu5Ac). The structures of complexes of HA of influenza A and B viruses with natural and synthetic sialic acid compounds were determined by X-ray crystallography [122–127]. The sialic acid-binding site is a shallow pocket located on the globular head of HA. Virus binding depends not only on HA affinity for the terminal sialic acid residues, but also on the structure of the underlying oligosaccharide and protein or lipid moieties of the receptors, as well as on the abundance and accessibility of receptors on the cell surface. Because of this complex mode of binding, the receptor-binding properties of influenza viruses can be affected by amino acid substitutions inside the sialic acid-binding pocket, on the pocket rim, and by distant mutations resulting in altered glycosylation or altered electrostatic charge of the globular head of HA (reviewed in ref. [128]). In natural glycoconjugates, sialic acids are α2-3- or α2-6-linked to Gal and GalNAc,

α2-6-linked to GlcNAc, or α2-8-linked to the second Sia residue. Influenza viruses generally do not bind to α2-8-linked Neu5Ac moieties and can recognize only α2-3- or α2-6-linked sialic acid epitopes (Neu5Acα2-3/6Gal, Neu5Acα2-3/6GalNAc, and Neu5Acα2-6GlcNAc).

Influenza virus host specificity can be explained in part by the difference in receptor-binding specificity for human and avian influenza viruses. Human influenza viruses preferentially bind to sialic acid by an α2,6-linkage (Neu5Acα2-6Gal) [129]. This preference is matched by Neu5Acα2-6Gal on epithelial cells in the human trachea. In contrast, avian influenza viruses preferentially recognize Neu5Acα2-3Gal that is matched by Neu5Acα2-3Gal on epithelial cells in the intestinal tract of waterfowl (the main replication site of avian influenza viruses). The receptor-binding specificity of human and avian influenza viruses suggests that avian influenza viruses need to acquire the ability to recognize human-type receptors to cause a pandemic. Indeed, the earliest isolates of the 1918, 1957, and 1968 pandemics possessed HA that, although of avian origin, recognized human-type receptors. In light of these findings, the infection of humans with highly pathogenic avian H5N1 viruses seemed to be surprising, particularly because the H5N1 viruses isolated from infected individuals in Hong Kong in 1997 preferentially recognized Neu5Acα2-3Gal [18]. Studies on human and avian virus infection in differentiated cultures of human airway epithelial cells (HAE) indicated, however, that some cells in the human airway epithelium express sufficient amounts of receptors to allow infection with avian viruses and that receptor specificity determines the viral cell tropism in the epithelium. Early in infection, human viruses preferentially infected non-ciliated cells, whereas avian viruses mainly infected ciliated cells [130, 131]. Other groups studied expression of viral receptors in human biopsies and archival tissues using lectins SNA, MAL-I, and/or MAL-II and human and avian influenza viruses as molecular probes [132–136]. The results obtained in these studies are consistent with the theory that paucity of receptors for avian viruses in the upper respiratory tract may be one of the factors preventing efficient human-to-human transmission.

The crystal structures of the avian HA complexes with the sialic acid compounds [124, 125] provided insights into the molecular basis of differential receptor specificity (Fig. 3). For H3 and H5 avian HAs, 3-linked sialyloligosaccharide is bound with the minimum-energy *trans* conformation of the Sia-Gal linkage. In this conformation, the glycosidic oxygen and the axial 4-OH group of Gal form hydrogen bonds with the side chain amide and carbonyl groups of the conserved glutamine in HA position 226 (226Q). The 6-linked sialyloligosaccharide LSTc binds to avian HAs in its own lowest energy *cis* conformation. In this case, the hydrophobic C6 methylene group of the α (2–6)-linkage projects downward toward the polar atoms of the side chain of 226Q. These unfavourable polar–non-polar interactions between 6-linked Gal and 226Q, account, at least in part, for the poor binding of avian viruses to 6-linked receptors. Thus, the receptor specificity of avian viruses is determined by the optimal fit to the HA receptor-binding site of the most abundant *trans* conformer of the Neu5Acα2-3Gal moiety and on a poor fit of the major *cis* conformer of the Neu5Acα2-6Gal epitope (for references see [137]).

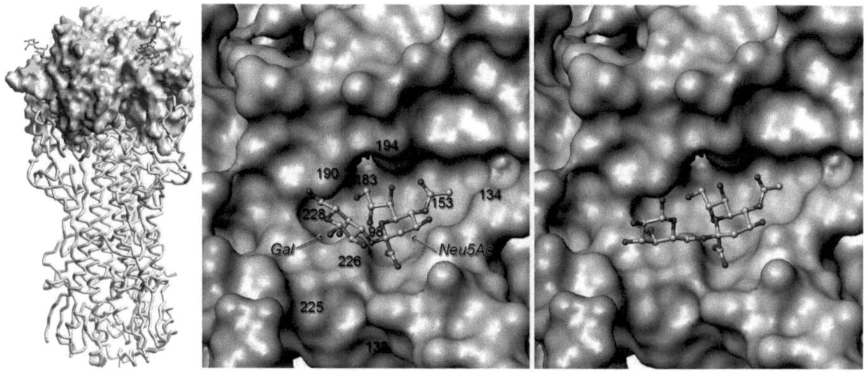

Fig. 3 Receptor-binding site of influenza virus HA. **Left image**. Individual monomers of the HA trimer are colored in white, pink and green. The HA globular head (amino acids 90-260) is shown as solvent accessible surface; amino acids conserved among avian viruses are colored yellow. Stick model illustrates position of bound sialyloligosaccharide LSTa (Neu5Acα2-3Galβ1-3GlcNAcβ1-3Galβ1-4Glc). **Center image** shows enlarged view of the RBS and Neu5Acα2-3Gal moiety of LSTa. Positions of some conserved amino acids of the avian RBS are indicated using H3 numbering system. **Right image** shows HA complex with Neu5Acα2-6Gal moiety of LSTc (Neu5Acα2-6Galβ1-4GlcNAcβ1-3Galβ1-4Glc). The figure is based on crystallographic data of Ha et al. for the HA of A/Duck/Ukraine/1/63 (H3N8) [125] (Courtesy M. Matrosovich, Marburg).

Six additional amino acids were found to be highly conserved in the receptor-binding site of most avian viruses (138A, 190E, 194L, 225G, 226Q, and 228G) when compared to the binding site of human viruses [138]. These amino acids either directly participate in hydrogen bond formation and van der Waals interactions with Neu5Acα2-3Gal moiety (190E, 194L, 225G, and 226Q), or make direct contacts with such amino acids (138A and 228G). Mutations in any of these positions were found to decrease the avian virus binding to the 3-linked galactose moiety and to lower the virus affinity for the Neu5Acα2-3Gal-containing sialylglycopolymers [139]. Thus, the conservation of these amino acids is required for the avian virus HA recognition of "the α2-3-linkage-specific motif" [124] and efficient virus attachment to target cells in birds.

8.4.2 Fusion

Cleavage of HA is necessary for membrane fusion which plays a central role in the initiation of infection [140, 141]. Virions are internalized by receptor mediated endocytosis, and the low pH within endosomes induces successive conformational changes of HA. The following steps can be discriminated in this process: (1) The cytoplasmic domain of HA2 dissociates from the matrix protein M1 after proton influx into the virion via M2 ion channels. (2) HA1 is folded back, and the N-terminal hydrophobic peptide (fusion peptide) of HA2 is released from a cavity in the stem region behind the cleavage site and is then immersed into the endosomal target membrane. (3) HA2 undergoes a drastic conformational change to a hairpin-like structure that merges lipids of virus and endosome [126]. After formation of fusion pores viral ribonucleoprotein complexes (RNPs) are delivered into the cytosol. RNPs are then transported into the nucleus, where transcription and replication occurs.

Proteolytic activation of HA is a prime determinant for the pathogenicity of avian influenza viruses. This concept was first derived from comparisons of naturally occurring strains [142–145] and was later further corroborated in numerous other studies (for review see [146–148]).

The HA of LPAI viruses is activated at the monobasic cleavage site, that is usually an arginine but sometimes a lysine [149, 150], by proteases secreted form epithelial cells that are present only in respiratory or intestinal tissues. Infection is therefore restricted to these organs. Little is known about the enzymes that activate these viruses in their natural setting. Recently, two serine proteases (TMPRSS2 and HAT) from human airway epithelium have been found to activate human influenza A viruses as well as a LPAI virus [151]. Bacterial proteases may also activate HAs of restricted cleavability and promote the development of pneumonia in mice after combined viral–bacterial infection [152].

HPAI viruses are activated by a different cleavage mechanism. Their HAs are activated at multibasic cleavage sites by furin, a member of the proprotein convertase family of eukaryotic subtilisin-like serine endoproteases [153]. The ubiquity of this enzyme accounts for the systemic infection typical for these viruses. Other proprotein convertases which resemble furin in structure and substrate specificity are PC1, PACE4, PC5/6, and LPC/PC7 [154]. Like furin, PC5/6 activates HAs with multibasic cleavage sites, whereas PC1, PACE, and LPC/PC7 do not [155, 156]. The HAs of most HPAI viruses have the consensus sequences R-X-K/R-R (for references see [147]) or R-X-X-R (for references see [157]) at the cleavage site, motifs that are both recognized by furin. Among the few exceptions to these rules are HAs which contain the unusual tetrapeptide K-K-K-R [158]. A multibasic cleavage site is not the only requirement for high cleavability. Another important determinant is a carbohydrate side chain close to the cleavage site that interferes with protease accessibility. Loss of this carbohydrate resulted in enhanced HA cleavability and pathogenicity [159].

HPAI viruses are derived from LPAI viruses by acquisition of a multibasic cleavage site. This phenomenon has been seen under experimental conditions as well as in natural outbreaks. It has first been observed in laboratory studies involving sequential cell culture passages of strains A/seal/Massachusetts/1/80 (H7N7) [160] and A/turkey/Oregon/71 (H7N3). In the latter case, the acquisition of the furin recognition motif resulted from recombination of the HA gene with 28S ribosomal RNA [92, 161]. The HA gene may not only recombine with cellular RNA but also with other viral gene segments, as has been observed recently when new HPAI viruses emerging in the field have been analysed. Thus, comparison of A/chicken/Chile/02 (H7N3) isolates revealed that the HA genes of the highly pathogenic strains had an insertion of 30 nucleotides at the cleavage site that was presumably derived from the nucleoprotein gene of the unrelated A/gull/Maryland/

704/77 (H13N6) virus [97]. Recombination between HA and matrix protein genes of the same virus generated the highly pathogenic A/chicken/BC/04 (H7N3) viruses [96]. Polymerase slippage has been suggested as an alternative strategy by which a multibasic cleavage site is generated [162, 163]. However, there are other examples where the mechanism of insertion is not clear [164]. In fact, it is generally not understood why insertion of a multibasic cleavage site is observed only with H5 and H7 haemagglutinins, but not with other serotypes.

8.5 Neuraminidase

NA is a type II membrane protein that is present in homotetrameric form in the viral envelope [165]. Each monomer consists of a cytoplasmic tail 6 amino acids in length, a stem region varying in length between 19 and 45 amino acids, and a carboxy-terminal globular head [166, 167]. The monomers are linked to dimers by disulfide bridges in the stalk region. Interspecies transmission of avian influenza viruses from aquatic birds to terrestrial poultry is often accompanied by a deletion in the stalk region and a reduced catalytic activity [18, 164]. The observation that the reduced catalytic activity of NA is compensated by mutations in HA resulting in decreased receptor affinity led to the concept that optimal virus replication depends on a balance between receptor binding by HA and receptor destruction by NA [168, 169].

The catalytic site of NA is located in the globular head region. It is in the centre of a propeller-like structure formed by four anti-parallel β-sheets [165]. *N*-Acetyl-D-neuraminic acid is bound by hydrogen bonds to amino acids R118, D151, R152, R224, E276, R292, and R371 (N2 numbering). The acetamido group is linked by van der Waals forces to W178 and I222. The amino acids directly interacting with sialic acid are stabilized by contacts with amino acids E119, R156, S179, D/N198, N294, and E425. All of these amino acids are conserved with the different NA subtypes.

NAs of avian strains have, in addition to the catalytic function, the capacity to bind erythrocytes [170–172]. NAs of human viruses are unable to haemadsorb. The haemadsorbing domain (HAD) is distinct from, but adjacent to, the catalytic site. It is formed by three amino acids loops containing S367, S370, S372 in the first, T400 and W403 in the second, and K432 in the third loop [170, 173, 174]. Recently, it could be shown that the haemadsorbing function enhances the catalytic activity of NA. This study revealed also that the haemadsorbing activity of NAs of early human isolates of the pandemics of 1918 and 1957 was reduced or completely absent. Thus, it appears that loss of the HAD is the result of an adaptive mutation involved in interspecies transmission from bird to man and has therefore to be considered as a pandemic marker [175].

The available evidence indicates that NA has several functions in the life cycle of influenza virus. It was Burnet who proposed more than 60 years ago that the RDE allows the virus to penetrate the mucus layer coating the respiratory epithelium and thus to infect its target cells [9]. This concept has recently been shown to be correct when it was found that the NA inhibitor oseltamivir prevented initiation of infection of human tracheo-bronchial cell cultures [176]. The second function of NA is at the end of the life cycle where it promotes virus release and clumping of virions by removing receptors from the cell surface and viral glycoproteins, respectively [177].

In influenza B virus, the NA gene encodes a second protein, NB. NB is also a membrane protein, but its function is not known.

8.6 *Influenza C Virus HEF*

Unlike influenza A and B viruses, influenza C virus has only one surface glycoprotein, HEF, which has receptor-destroying, receptor-binding, and fusion activity. HEF binds specifically to 9-*0*-acetyl, 5-*N*-acetylneuraminic acids, and the receptor destroying activity is that of a 9-*0*-acetyl esterase [178]. All three functions of HEF are segregated into structurally distinct domains. Although HEF and HA show only 12% sequence identity, both proteins are structurally similar [179].

8.7 *Matrix Protein M1*

M1 is the most abundant virion protein. It lines the inner surface of the lipid bilayer and plays an essential role in virus budding. It also regulates the movement of RNPs out of the nucleus and inhibits viral RNA synthesis at late stages of viral replication [12].

8.8 *M2 and BM2*

M2 of influenza A viruses is a tetrameric type III membrane protein that has ion channel activity. It functions during virus entry where it is responsible for acidifying the core of the virus particle which triggers dissociation of M1 from the viral RNP (uncoating). With HPAI viruses, M2 acts also at a later stage in infection by reducing the pH of the trans Golgi network to protect HA from premature conformational change [180]. The BM2 protein of influenza B virus plays a similar role in entry as M2.

8.9 *Non-structural Protein NS1*

NS1 is a small protein with distinct functional domains that regulate a number of critical events during influenza virus replication. It enhances virus replication by activating the cellular enzyme phophatidylinositol-3-kinase and by down-regulating apoptosis. It also forms complexes with eukaryotic translation initiation factors and enhances initiation of viral mRNA translation. Furthermore, it suppresses the induction of RNA interference and interferes with cellular mRNA processing. The most prominent role of NS1 is that of an interferon (IFN) antagonist. IFN antagonism is accomplished by several mechanisms. These include inactivation of transcription factor IRF-3 which is crucial for IFN gene expression, inhibition of transcription of IFN and IFN-regulated genes by blocking mRNA processing, and down-regulation of IFN response by blocking nuclear export of cellular mRNA (for references see [181]). Several amino acid exchanges of NS1 and deletions at the C-terminus have been associated with differences in pathogenicity, but the underlying mechanisms are not clear. Interestingly, the swine origin 2009 influenza virus resembles in these NS1 markers viruses of low pathogenicity [182].

8.10 Nuclear Export Protein NEP/NS2

This protein mediates the nuclear export of newly synthesized RNPs. It is also present in small amounts in virus particles.

9 Prophylaxis and Therapy

9.1 Vaccines

Inactivated influenza A and B virus vaccines are most widely used in humans. The vaccine is reformulated each year to include the strains thought most likely to be prevailing. The vaccine virus is mostly still grown in embryonated eggs. Because many human influenza virus isolates do not grow to high yield in eggs, a reassortment virus is made using the high egg-yielding PR/8/34 genetic backbone and the HA and NA genes of the candidate virus. Reverse genetics procedures speed up the time needed to produce a high-yielding reassortment over the former method of mixed infection and selecting the virus from random plaques. One egg yields one to three doses of vaccine. The vaccine is administered intramuscularly. Cell culture-based vaccines are also developed. Vaccination with inactivated virus has been shown to consistently confer resistance to illness (reduced frequency and severity of disease) and to protect to a somewhat lesser extent against infection with influenza A and B viruses.

Live attenuated vaccines are reassortants of the genetic backbone of cold-adapted viruses and the current HA and NA genes. The cold-adapted virus replicates efficiently in the nasopharynx to induce prospective immunity. However, replication is restricted at higher temperatures, including those present in the lower airways and lungs. In clinical studies with matched strains, live virus vaccines demonstrated 87% efficacy in children and 85% in adults. There is continued interest in the use of a live influenza virus vaccine because infection of the respiratory tract stimulates both systemic and local immunity, and in principle should stimulate cell-mediated immunity. Thus, all components of the human immune response are brought into action. Furthermore, there is added advantage of acceptance of a nasal spray rather than a needle injection in young children.

9.2 Antivirals

Two classes of specific anti-influenza drugs have been developed to date: the aminoadamantanes: amantadine and rimantadine, developed in the early 1960s and targeted against the M2 proton channel of influenza A virus; and the more recently developed inhibitors of the viral NA: zanamivir and oseltamivir, further discussed in other chapters. Prior to the late 1990s application of amantadine and rimantadine was limited since the drugs are not effective against influenza B. There were also concerns about effectiveness and central nervous system complications. The propensity for resistance emergence particularly among the currently circulating H3N2, H1N1, and H5N1 viruses and a reduced enthusiasm for widespread use in a seasonal context have led to specific recommendations against the use of this class of drug against seasonal influenza. In contrast, resistance to the NA inhibitors was believed to emerge less readily, especially since some of the more frequent mutations substantially compromise virus infectivity and transmission. However, with the spectre of the next pandemic looming and recent examples of the more frequent emergence of oseltamivir resistance, the impact on the effectiveness of drug treatment of mutations which reduce drug susceptibility as well as those which cause high resistance is of major concern, especially in relation to drug stockpiled as the first line of defence to counter a pandemic [180].

Acknowledgements I gratefully acknowledge the editorial help of Sabine Fischbach.

References

1. Shope RE (1931) Swine influenza. 1. Experimental transmission and pathology. J Exp Med 54:349–360
2. Smith W, Andrewes CH, Laidlaw PP (1933) A virus obtained from influenza patients. Lancet 66–68
3. Centanni E (1902) Die Vogelpest. Beitrag zu dem durch Kerzen filtrierbaren Virus (Fowl plague. Report on the candle filterable virus). *Centralblatt für Bakteriologie*. Parasitenkunde und Infektionskrankheiten I Abteilung: medizinische-hygienische Bakteriologie und tierische Parasitenkunde 31:145–152
4. Lode A, Gruber F (1901) Bakteriologische Studien über die Aetiologie einer epidemischen Erkrankung der Hühner in Tirol. Centralblatt für Bakteriologie, Parasitenkunde und

Infektionskrankheiten I Abteilung: medizinische-hygienische Bakteriologie und tierische Parasitenkunde 30:593–604

5. Schaefer W (1955) Vergleichende sero-immunologische Untersuchungen über die Viren der Influenza und klassischen Geflügelpest (Comparative sero-immunological investigations on the viruses of influenza and classical fowl plague). Z Naturforsch 10b:81–91
6. Webster RG, Bean WJ, Gorman OT, Chambers TM, Kawaoka Y (1992) Evolution and ecology of influenza A viruses. Microbiol Rev 56:152–179
7. Hirst GK (1941) The agglutination of red cells by allantoic fluid of chick embryos infected with influenza virus. Science 94:22–23
8. McClelland L, Hare R (1941) The adsorption of influenza virus by red cells and a new in vitro method of measuring antibodies for influenza virus. Can J Public Health 32:530–538
9. Burnet FM, Stone JD (1947) The receptor destroying enzyme of *V. cholerae*. Aust J Exp Biol Med Sci 25:227–233
10. Gottschalk A, Lind PE (1949) Product of interaction between influenza virus enzyme and ovomucin. Nature 164:232
11. Klenk E, Faillard H, Lempfrid H (1955) Über die enzymatische Wirkung von Influenza Virus. Z Physiol Chem 301:235–246
12. Shaw ML, Palese P (2008) Orthomyxoviruses: molecular biology. In: Mahy BWJ, Van Regenmortel M (eds) Encyclopedia of virology, vol 3. Elsevier, Amsterdam, pp 483–489
13. Olsen B, Munster VJ, Wallensten A, Waldenstrom J, Osterhaus AD, Fouchier RA (2006) Global patterns of influenza a virus in wild birds. Science 312:384–388
14. Alexander DJ, Brown IH (2009) History of highly pathogenic avian influenza. Rev Sci Tech 28:19–38
15. Campitelli L, Mogavero E, De Marco MA, Delogu M, Puzelli S, Frezza F, Facchini M, Chiapponi C, Foni E, Cordioli P et al (2004) Interspecies transmission of an H7N3 influenza virus from wild birds to intensively reared domestic poultry in Italy. Virology 323:24–36
16. Roehm C, Horimoto T, Kawaoka Y, Suss J, Webster RG (1995) Do hemagglutinin genes of highly pathogenic avian influenza viruses constitute unique phylogenetic lineages? Virology 209:664–670
17. Chen H, Smith GJ, Li KS, Wang J, Fan XH, Rayner JM, Vijaykrishna D, Zhang JX, Zhang LJ, Guo CT et al (2006) Establishment of multiple sublineages of H5N1 influenza virus in Asia: implications for pandemic control. Proc Natl Acad Sci USA 103:2845–2850
18. Matrosovich M, Zhou N, Kawaoka Y, Webster R (1999) The surface glycoproteins of H5 influenza viruses isolated from humans, chickens, and wild aquatic birds have distinguishable properties. J Virol 73:1146–1155
19. Matrosovich MN, Krauss S, Webster RG (2001) H9N2 Influenza A Viruses from Poultry in Asia Have Human Virus-like Receptor Specificity. Virology 281:156–162
20. Horimoto T, Kawaoka Y (2005) Influenza: lessons from past pandemics, warnings from current incidents. Nat Rev Microbiol 3:591–600
21. Gabriel G, Dauber B, Wolff T, Planz O, Klenk HD, Stech J (2005) The viral polymerase mediates adaptation of an avian influenza virus to a mammalian host. Proc Natl Acad Sci USA 102:18590–18595
22. Ludwig S, Stitz L, Planz O, Van H, Fitch WM, Scholtissek C (1995) European swine virus as a possible source for the next influenza pandemic? Virology 212:555–561
23. Scholtissek C (1990) Pigs as "mixing vessels" for the creation of new pandemic influenza A viruses. Med Principles Pract 2:65–71
24. Taubenberger JK, Morens DM (2009) Pandemic influenza–including a risk assessment of H5N1. Rev Sci Tech 28:187–202
25. Taubenberger JK, Reid AH, Krafft AE, Bijwaard KE, Fanning TG (1997) Initial genetic characterization of the 1918 "Spanish" influenza virus. Science 275:1793–1796
26. Reid AH, Fanning TG, Hultin JV, Taubenberger JK (1999) Origin and evolution of the 1918 "Spanish" influenza virus hemagglutinin gene. Proc Natl Acad Sci USA 96:1651–1656
27. Tumpey TM, Basler CF, Aguilar PV, Zeng H, Solorzano A, Swayne DE, Cox NJ, Katz JM, Taubenberger JK, Palese P, Garcia-Sastre A (2005) Characterization of the reconstructed 1918 Spanish influenza pandemic virus. Science 310:77–80

28. Kobasa D, Takada A, Shinya K, Hatta M, Halfmann P, Theriault S, Suzuki H, Nishimura H, Mitamura K, Sugaya N et al (2004) Enhanced virulence of influenza A viruses with the hemagglutinin of the 1918 pandemic virus. Nature 431:703–707
29. Tumpey TM, Garcia-Sastre A, Mikulasova A, Taubenberger JK, Swayne DE, Palese P, Basler CF (2002) Existing antivirals are effective against influenza viruses with genes from the 1918 pandemic virus. Proc Natl Acad Sci USA 99:13849–13854
30. Tumpey TM, Garcia-Sastre A, Taubenberger JK, Palese P, Swayne DE, Basler CF (2004) Pathogenicity and immunogenicity of influenza viruses with genes from the 1918 pandemic virus. Proc Natl Acad Sci USA 101:3166–3171
31. Van Hoeven N, Pappas C, Belser JA, Maines TR, Zeng H, Garcia-Sastre A, Sasisekharan R, Katz JM, Tumpey TM (2009) Human HA and polymerase subunit PB2 proteins confer transmission of an avian influenza virus through the air. Proc Natl Acad Sci USA 106: 3366–3371
32. Watanabe T, Watanabe S, Shinya K, Kim JH, Hatta M, Kawaoka Y (2009) Viral RNA polymerase complex promotes optimal growth of 1918 virus in the lower respiratory tract of ferrets. Proc Natl Acad Sci USA 106:588–592
33. Geiss GK, Salvatore M, Tumpey TM, Carter VS, Wang X, Basler CF, Taubenberger JK, Bumgarner RE, Palese P, Katze MG, Garcia-Sastre A (2002) Cellular transcriptional profiling in influenza A virus-infected lung epithelial cells: the role of the nonstructural NS1 protein in the evasion of the host innate defense and its potential contribution to pandemic influenza. Proc Natl Acad Sci USA 99:10736–10741
34. McAuley JL, Hornung F, Boyd KL, Smith AM, McKeon R, Bennink J, Yewdell JW, McCullers JA (2007) Expression of the 1918 influenza A virus PB1-F2 enhances the pathogenesis of viral and secondary bacterial pneumonia. Cell Host Microbe 2:240–249
35. Kash JC, Tumpey TM, Proll SC, Carter V, Perwitasari O, Thomas MJ, Basler CF, Palese P, Taubenberger JK, Garcia-Sastre A et al (2006) Genomic analysis of increased host immune and cell death responses induced by 1918 influenza virus. Nature 443:578–581
36. Kobasa D, Jones SM, Shinya K, Kash JC, Copps J, Ebihara H, Hatta Y, Kim JH, Halfmann P, Hatta M et al (2007) Aberrant innate immune response in lethal infection of macaques with the 1918 influenza virus. Nature 445:319–323
37. Kawaoka Y, Krauss S, Webster RG (1989) Avian-to-human transmission of the PB1 gene of influenza A viruses in the 1957 and 1968 pandemics. J Virol 63:4603–4608
38. Scholtissek C, Rohde W, Von Hoyningen V, Rott R (1978) On the origin of the human influenza virus subtypes H2N2 and H3N2. Virology 87:13–20
39. Nakajima K, Desselberger U, Palese P (1978) Recent human influenza A (H1N1) viruses are closely related genetically to strains isolated in 1950. Nature 274:334–339
40. Scholtissek C, von Hoyningen V, Rott R (1978) Genetic relatedness between the new 1977 epidemic strains (H1N1) of influenza and human influenza strains isolated between 1947 and 1957 (H1N1). Virology 89:613–617
41. Holmes EC, Ghedin E, Miller N, Taylor J, Bao Y, St George K, Grenfell BT, Salzberg SL, Fraser CM, Lipman DJ, Taubenberger JK (2005) Whole-genome analysis of human influenza A virus reveals multiple persistent lineages and reassortment among recent H3N2 viruses. PLoS Biol 3:e300
42. Kilbourne ED (1977) Influenza pandemics in perspective. JAMA 237:1225–1228
43. Pyle GF (1986) The diffusion of influenza: patterns and paradigms. Rowan & Littlefield, New Jersey
44. Noble GR (1982) Epidemiological and clinical aspects of influenza. In: Beare AS (ed) Basic and applied influenza research. CRC Press, Boca Raton, FL, pp 11–50
45. Kilbourne ED, Smith C, Brett I, Pokorny BA, Johansson B, Cox N (2002) The total influenza vaccine failure of 1947 revisited: major intrasubtypic antigenic change can explain failure of vaccine in a post-World War II epidemic. Proc Natl Acad Sci USA 99:10748–10752
46. Garten RJ, Davis CT, Russell CA, Shu B, Lindstrom S, Balish A, Sessions WM, Xu X, Skepner E, Deyde V et al (2009) Antigenic and genetic characteristics of swine-origin 2009 A (H1N1) influenza viruses circulating in humans. Science 325:197–201

47. Claas EC, Osterhaus AD, van Beek R, De Jong JC, Rimmelzwaan GF, Senne DA, Krauss S, Shortridge KF, Webster RG (1998) Human influenza A H5N1 virus related to a highly pathogenic avian influenza virus. Lancet 351:472–477
48. Subbarao K, Klimov A, Katz J, Regnery H, Lim W, Hall H, Perdue M, Swayne D, Bender C, Huang J et al (1998) Characterization of an avian influenza A (H5N1) virus isolated from a child with a fatal respiratory illness. Science 279:393–396
49. Xu X, Subbarao CNJ, Guo Y (1999) Genetic characterization of the pathogenic influenza A/Goose/Guangdong/1/96 (H5N1) virus: similarity of its hemagglutinin gene to those of H5N1 viruses from the 1997 outbreaks in Hong Kong. Virology 261:15–19
50. Cauthen AN, Swayne DE, Schultz-Cherry S, Perdue ML, Suarez DL (2000) Continued circulation in China of highly pathogenic avian influenza viruses encoding the hemagglutinin gene associated with the 1997 H5N1 outbreak in poultry and humans. J Virol 74:6592–6599
51. Guan Y, Peiris JS, Lipatov AS, Ellis TM, Dyrting KC, Krauss S, Zhang LJ, Webster RG, Shortridge KF (2002) Emergence of multiple genotypes of H5N1 avian influenza viruses in Hong Kong SAR. Proc Natl Acad Sci USA 99:8950–8955
52. World Health Organization (2003) WHO: disease alert: 10 December 2003. Influenza A (H9N2) in Hong Kong Special Administrative Region of China
53. Li KS, Guan Y, Wang J, Smith GJ, Xu KM, Duan L, Rahardjo AP, Puthavathana P, Buranathai C, Nguyen TD et al (2004) Genesis of a highly pathogenic and potentially pandemic H5N1 influenza virus in eastern Asia. Nature 430:209–213
54. Liu J, Xiao H, Lei F, Zhu Q, Qin K, Zhang XW, Zhang XL, Zhao D, Wang G, Feng Y et al (2005) Highly pathogenic H5N1 influenza virus infection in migratory birds. Science 309:1206
55. Kuiken T, Rimmelzwaan G, van Riel D, van Amerongen G, Baars M, Fouchier R, Osterhaus A (2004) Avian H5N1 influenza in cats. Science 306:241
56. Enserink M, Kaiser J (2004) Virology. Avian flu finds new mammal hosts. Science 305:1385
57. Weber S, Harder T, Starick E, Beer M, Werner O, Hoffmann B, Mettenleiter TC, Mundt E (2007) Molecular analysis of highly pathogenic avian influenza virus of subtype H5N1 isolated from wild birds and mammals in northern Germany. J Gen Virol 88:554–558
58. de Jong MD, Simmons CP, Thanh TT, Hien VM, Smith GJ, Chau TN, Hoang DM, Chau NV, Khanh TH, Dong VC et al (2006) Fatal outcome of human influenza A (H5N1) is associated with high viral load and hypercytokinemia. Nat Med 12:1203–1207
59. Banks J, Speidel E, Alexander DJ (1998) Characterisation of an avian influenza A virus isolated from a human–is an intermediate host necessary for the emergence of pandemic influenza viruses? Arch Virol 143:781–787
60. Kurtz J, Manvell RJ, Banks J (1996) Avian influenza virus isolated from a woman with conjunctivitis. Lancet 348:901–902
61. Fouchier RA, Schneeberger PM, Rozendaal FW, Broekman JM, Kemink SA, Munster V, Kuiken T, Rimmelzwaan GF, Schutten M, Van Doornum GJ et al (2004) Avian influenza A virus (H7N7) associated with human conjunctivitis and a fatal case of acute respiratory distress syndrome. Proc Natl Acad Sci USA 101:1356–1361
62. Koopmans M, Vennema H, Heersma H, van Strien E, van Duynhoven Y, Brown D, Reacher M, Lopman B (2003) Early identification of common-source foodborne virus outbreaks in Europe. Emerg Infect Dis 9:1136–1142
63. van Kolfschooten F (2003) Dutch veterinarian becomes first victim of avian influenza. Lancet 361:1444
64. Bosman A (2004) Avian flu epidemic 2003. RIVM report
65. Guo YJ, Krauss S, Senne DA, Mo IP, Lo KS, Xiong XP, Norwood M, Shortridge KF, Webster RG, Guan Y (2000) Characterization of the pathogenicity of members of the newly established H9N2 influenza virus lineages in Asia. Virology 267:279–288
66. Peiris M, Yuen KY, Leung CW, Chan KH, Ip PL, Lai RW, Orr WK, Shortridge KF (1999) Human infection with influenza H9N2. Lancet 354:916–917

67. Lin YP, Shaw M, Gregory V, Cameron K, Lim W, Klimov A, Subbarao K, Guan Y, Krauss S, Shortridge K et al (2000) Avian-to-human transmission of H9N2 subtype influenza A viruses: relationship between H9N2 and H5N1 human isolates. Proc Natl Acad Sci USA 97:9654–9658
68. Choi YK, Ozaki H, Webby RJ, Webster RG, Peiris JS, Poon L, Butt C, Leung YH, Guan Y (2004) Continuing evolution of H9N2 influenza viruses in Southeastern China. J Virol 78:8609–8614
69. Gratzl E, Koehler H (1968) Gefluegelpest Spezielle Pathologie und Therapie der Gefluegelkrankheiten. Ferdinand Enke Verlag, Stuttgart, Germany
70. Narayan O, Thorsen J, Hulland TJ, Ankeli G, Joseph PG (1972) Pathogenesis of lethal influenza virus infection in turkeys. I. Extraneural phase of infection. J Comp Pathol 82:129–137
71. Van Campen H, Easterday BC, Hinshaw VS (1989) Virulent avian influenza A viruses: their effect on avian lymphocytes and macrophages in vivo and in vitro. J Gen Virol 70(Pt 11):2887–2895
72. Van Campen H, Easterday BC, Hinshaw VS (1989) Destruction of lymphocytes by a virulent avian influenza A virus. J Gen Virol 70(Pt 2):467–472
73. Kobayashi Y, Horimoto T, Kawaoka Y, Alexander DJ, Itakura C (1996) Pathological studies of chickens experimentally infected with two highly pathogenic avian influenza viruses. Avian Pathol 25:285–304
74. Feldmann A, Schafer MK, Garten W, Klenk HD (2000) Targeted infection of endothelial cells by avian influenza virus A/FPV/Rostock/34 (H7N1) in chicken embryos. J Virol 74:8018–8027
75. Area E, Martin-Benito J, Gastaminza P, Torreira E, Valpuesta JM, Carrascosa JL, Ortin J (2004) 3D structure of the influenza virus polymerase complex: localization of subunit domains. Proc Natl Acad Sci USA 101:308–313
76. Fodor E, Smith M (2004) The PA subunit is required for efficient nuclear accumulation of the PB1 subunit of the influenza A virus RNA polymerase complex. J Virol 78:9144–9153
77. Deng T, Sharps JL, Brownlee GG (2006) Role of the influenza virus heterotrimeric RNA polymerase complex in the initiation of replication. J Gen Virol 87:3373–3377
78. Engelhardt OG, Smith M, Fodor E (2005) Association of the influenza A virus RNA-dependent RNA polymerase with cellular RNA polymerase II. J Virol 79:5812–5818
79. Ye Q, Krug RM, Tao YJ (2006) The mechanism by which influenza A virus nucleoprotein forms oligomers and binds RNA. Nature 444:1078–1082
80. Hay AJ, Lomniczi B, Bellamy AR, Skehel JJ (1977) Transcription of the influenza virus genome. Virology 83:337–355
81. Ulmanen I, Broni BA, Krug RM (1981) Role of two of the influenza virus core P proteins in recognizing cap 1 structures (m7G pppNm) on RNAs and in initiating viral RNA transcription. Proc Natl Acad Sci USA 78:7355–7359
82. Shi L, Summers DF, Peng Q, Galarz JM (1995) Influenza A virus RNA polymerase subunit PB2 is the endonuclease which cleaves host cell mRNA and functions only as the trimeric enzyme. Virology 208:38–47
83. Poon LL, Pritlove DC, Fodor E, Brownlee GG (1999) Direct evidence that the poly(A) tail of influenza A virus mRNA is synthesized by reiterative copying of a U track in the virion RNA template. J Virol 73:3473–3476
84. Zheng H, Lee HA, Palese P, Garcia-Sastre A (1999) Influenza A virus RNA polymerase has the ability to stutter at the polyadenylation site of a viral RNA template during RNA replication. J Virol 73:5240–5243
85. Poon LL, Pritlove DC, Sharps J, Brownlee GG (1998) The RNA polymerase of influenza virus, bound to the 5' end of virion RNA, acts in cis to polyadenylate mRNA. J Virol 72:8214–8219
86. Hay AJ, Skehel JJ, McCauley J (1982) Characterization of influenza virus RNA complete transcripts. Virology 116:517–522

87. Castrucci MR, Kawaoka Y (1993) Biologic importance of neuraminidase stalk length in influenza A virus. J Virol 67:759–764
88. Deshpande KL, Naeve CW, Webster RG (1985) The neuraminidases of the virulent and avirulent A/Chicken/Pennsylvania/83 (H5N2) influenza A viruses: sequence and antigenic analyses. Virology 147:49–60
89. Els MC, Air GM, Murti KG, Webster RG, Laver WG (1985) An 18-amino acid deletion in an influenza neuraminidase. Virology 142:241–247
90. Luo G, Chung J, Palese P (1993) Alterations of the stalk of the influenza virus neuraminidase: deletions and insertions. Virus Res 29:141–153
91. Steuler H, Rohde W, Scholtissek C (1984) Sequence of the neuraminidase gene of an avian influenza A virus (A/parrot/ulster/73, H7N1). Virology 135:118–124
92. Khatchikian D, Orlich M, Rott R (1989) Increased viral pathogenicity after insertion of a 28S ribosomal RNA sequence into the hemagglutinin gene of an influenza virus. Nature 340:156–157
93. Orlich M, Khatchikian D, Teigler A, Rott R (1990) Structural variation occurring in the hemagglutinin of influenza virus A/turkey/Oregon/71 during adaptation to different cell types. Virology 176:531–538
94. Bergmann M, Garcia-Sastre A, Palese P (1992) Transfection-mediated recombination of influenza A virus. J Virol 66:7576–7580
95. Orlich M, Gottwald H, Rott R (1994) Nonhomologous recombination between the hemagglutinin gene and the nucleoprotein gene of an influenza virus. Virology 204:462–465
96. Pasick J, Handel K, Robinson J, Copps J, Ridd D, Hills K, Kehler H, Cottam-Birt C, Neufeld J, Berhane Y, Czub S (2005) Intersegmental recombination between the hemagglutinin and matrix genes was responsible for the emergence of a highly pathogenic H7N3 avian influenza virus in British Columbia. J Gen Virol 86:727–731
97. Suarez DL, Senne DA, Banks J, Brown IH, Essen SC, Lee CW, Manvell RJ, Mathieu-Benson C, Moreno V, Pedersen JC et al (2004) Recombination resulting in virulence shift in avian influenza outbreak, Chile. Emerg Infect Dis 10:693–699
98. Subbarao EK, London W, Murphy BR (1993) A single amino acid in the PB2 gene of influenza A virus is a determinant of host range. J Virol 67:1761–1764
99. Massin P, van der WS, Naffakh N (2001) Residue 627 of PB2 is a determinant of cold sensitivity in RNA replication of avian influenza viruses. J Virol 75:5398–5404
100. Gao P, Watanabe S, Ito T, Goto H, Wells K, McGregor M, Cooley AJ, Kawaoka Y (1999) Biological heterogeneity, including systemic replication in mice, of H5N1 influenza A virus isolates from humans in Hong Kong. J Virol 73:3184–3189
101. Hatta M, Gao P, Halfmann P, Kawaoka Y (2001) Molecular basis for high virulence of Hong Kong H5N1 influenza A viruses. Science 293:1840–1842
102. Salomon R, Franks J, Govorkova EA, Ilyushina NA, Yen HL, Hulse-Post DJ, Humberd J, Trichet M, Rehg JE, Webby RJ et al (2006) The polymerase complex genes contribute to the high virulence of the human H5N1 influenza virus isolate A/Vietnam/1203/04. J Exp Med 203:689–697
103. Shinya K, Hamm S, Hatta M, Ito H, Ito T, Kawaoka Y (2004) PB2 amino acid at position 627 affects replicative efficiency, but not cell tropism, of Hong Kong H5N1 influenza A viruses in mice. Virology 320:258–266
104. Govorkova EA, Rehg JE, Krauss S, Yen HL, Guan Y, Peiris M, Nguyen TD, Hanh TH, Puthavathana P, Long HT et al (2005) Lethality to ferrets of H5N1 influenza viruses isolated from humans and poultry in 2004. J Virol 79:2191–2198
105. Puthavathana P, Auewarakul P, Charoenying PC, Sangsiriwut K, Pooruk P, Boonnak K, Khanyok R, Thawachsupa P, Kijphati R, Sawanpanyalert P (2005) Molecular characterization of the complete genome of human influenza H5N1 virus isolates from Thailand. J Gen Virol 86:423–433
106. Taubenberger JK, Reid AH, Lourens RM, Wang R, Jin G, Fanning TG (2005) Characterization of the 1918 influenza virus polymerase genes. Nature 437:889–893

107. Li Z, Chen H, Jiao P, Deng G, Tian G, Li Y, Hoffmann E, Webster RG, Matsuoka Y, Yu K (2005) Molecular basis of replication of duck H5N1 influenza viruses in a mammalian mouse model. J Virol 79:12058–12064
108. Gabriel G, Abram M, Keiner B, Wagner R, Klenk HD, Stech J (2007) Differential polymerase activity in avian and mammalian cells determines host range of influenza virus. J Virol 81:9601–9604
109. Naffakh N, Massin P, Escriou N, Crescenzo-Chaigne B, van der WS (2000) Genetic analysis of the compatibility between polymerase proteins from human and avian strains of influenza A viruses. J GenVirol 81:1283–1291
110. Gabriel G, Herwig A, Klenk HD (2008) Interaction of polymerase subunit PB2 and NP with importin alpha1 is a determinant of host range of influenza A virus. PLoS Pathog 4:e11
111. Tarendeau F, Boudet J, Guilligay D, Mas PJ, Bougault CM, Boulo S, Baudin F, Ruigrok RW, Daigle N, Ellenberg J et al (2007) Structure and nuclear import function of the C-terminal domain of influenza virus polymerase PB2 subunit. Nat Struct Mol Biol 14:229–233
112. Tarendeau F, Crepin T, Guilligay D, Ruigrok RW, Cusack S, Hart DJ (2008) Host determinant residue lysine 627 lies on the surface of a discrete, folded domain of influenza virus polymerase PB2 subunit. PLoS Pathog 4:e1000136
113. Mayer D, Molawi K, Martinez-Sobrido L, Ghanem A, Thomas S, Baginsky S, Grossmann J, Garcia-Sastre A, Schwemmle M (2007) Identification of cellular interaction partners of the influenza virus ribonucleoprotein complex and polymerase complex using proteomic-based approaches. J Proteome Res 6:672–682
114. Chen W, Calvo PA, Malide D, Gibbs J, Schubert U, Bacik I, Basta S, O'Neill R, Schickli J, Palese P et al (2001) A novel influenza A virus mitochondrial protein that induces cell death. Nat Med 7:1306–1312
115. Chang S, Zhang J, Liao X, Zhu X, Wang D, Zhu J, Feng T, Zhu B, Gao GF, Wang J et al (2007) Influenza Virus Database (IVDB): an integrated information resource and analysis platform for influenza virus research. Nucleic Acids Res 35:D376–380
116. Russell RJ, Gamblin SJ, Haire LF, Stevens DJ, Xiao B, Ha Y, Skehel JJ (2004) H1 and H7 influenza hemagglutinin structures extend a structural classification of hemagglutinin subtypes. Virology 325:287–296
117. Stevens J, Corper AL, Basler CF, Taubenberger JK, Palese P, Wilson IA (2004) Structure of the uncleaved human H1 hemagglutinin from the extinct 1918 influenza virus. Science 303:1866–1870
118. Chen J, Lee KH, Steinhauer DA, Stevens DJ, Skehel JJ, Wiley DC (1998) Structure of the hemagglutinin precursor cleavage site, a determinant of influenza pathogenicity and the origin of the labile conformation. Cell 95:409–417
119. Wilson IA, Skehel JJ, Wiley DC (1981) Structure of the hemagglutinin membrane glycoprotein of influenza virus at 3 A resolution. Nature 289:366–373
120. Ha Y, Stevens DJ, Skehel JJ, Wiley DC (2002) H5 avian and H9 swine influenza virus hemagglutinin structures: possible origin of influenza subtypes. Embo J 21:865–875
121. Stevens J, Blixt O, Tumpey TM, Taubenberger JK, Paulson JC, Wilson IA (2006) Structure and receptor specificity of the hemagglutinin from an H5N1 influenza virus. Science 312:404–410
122. Eisen MB, Sabesan S, Skehel JJ, Wiley DC (1997) Binding of the influenza A virus to cell-surface receptors: structures of five hemagglutinin-sialyloligosaccharide complexes determined by X-ray crystallography. Virology 232:19–31
123. Gamblin SJ, Haire LF, Russell RJ, Stevens DJ, Xiao B, Ha Y, Vasisht N, Steinhauer DA, Daniels RS, Elliot A et al (2004) The structure and receptor-binding properties of the 1918 influenza hemagglutinin. Science 303:1838–1842
124. Ha Y, Stevens DJ, Skehel JJ, Wiley DC (2001) X-ray structures of H5 avian and H9 swine influenza virus hemagglutinins bound to avian and human receptor analogs. Proc Natl Acad Sci USA 98:11181–11186

125. Ha Y, Stevens DJ, Skehel JJ, Wiley DC (2003) X-ray structure of the hemagglutinin of a potential H3 avian progenitor of the 1968 Hong Kong pandemic influenza virus. Virology 309:209–218
126. Skehel JJ, Wiley DC (2000) Receptor binding and membrane fusion in virus entry: the influenza hemagglutinin. Annu Rev Biochem 69:531–569
127. Wang Q, Tian X, Chen X, Ma J (2007) Structural basis for receptor specificity of influenza B virus hemagglutinin. Proc Natl Acad Sci USA 104:16874–16879
128. Matrosovich MN, Klenk HD, Kawaoka Y (2006) Receptor specificity, host-range and pathogenicity of influenza viruses. In: Kawaoka Y (ed) Influenza virology: current topics. Horizon Press, Wymondham, pp 95–137
129. Rogers GN, Paulson JC (1983) Receptor determinants of human and animal influenza virus isolates: differences in receptor specificity of the H3 hemagglutinin based on species of origin. Virology 127:361–373
130. Matrosovich M, Matrosovich T, Uhlendorff J, Garten W, Klenk HD (2007) Avian-virus-like receptor specificity of the hemagglutinin impedes influenza virus replication in cultures of human airway epithelium. Virology 361:384–390
131. Matrosovich MN, Matrosovich TY, Gray T, Roberts NA, Klenk HD (2004) Human and avian influenza viruses target different cell types in cultures of human airway epithelium. Proc Natl Acad Sci USA 101:4620–4624
132. Nicholls JM, Chan MC, Chan WY, Wong HK, Cheung CY, Kwong DL, Wong MP, Chui WH, Poon LL, Tsao SW et al (2007) Tropism of avian influenza A (H5N1) in the upper and lower respiratory tract. Nat Med 13:147–149
133. Shinya K, Ebina M, Yamada S, Ono M, Kasai N, Kawaoka Y (2006) Avian flu: influenza virus receptors in the human airway. Nature 440:435–436
134. van Riel D, Munster VJ, de Wit E, Rimmelzwaan GF, Fouchier RA, Osterhaus AD, Kuiken T (2006) H5N1 virus attachment to lower respiratory tract. Science 312:399
135. van Riel D, Munster VJ, de Wit E, Rimmelzwaan GF, Fouchier RA, Osterhaus AD, Kuiken T (2007) Human and avian influenza viruses target different cells in the lower respiratory tract of humans and other mammals. Am J Pathol 171:1215–1223
136. Yao L, Korteweg C, Hsueh W, Gu J (2007) Avian influenza receptor expression in H5N1-infected and noninfected human tissues. FASEB J. doi:DOI:10.1096/fj.1006-7880com
137. Matrosovich MN, Gambarian AS, Klenk HD (2008) Receptor specificity of influenza viruses and its alteration during interspecies transmission. In: Klenk HD, Matrosovich MN, Stech J (eds) Avian influenza, vol 27. Karger, Basel, pp 134–155
138. Matrosovich MN, Gambaryan AS, Teneberg S, Piskarev VE, Yamnikova SS, Lvov DK, Robertson JS, Karlsson KA (1997) Avian influenza A viruses differ from human viruses by recognition of sialyloligosaccharides and gangliosides and by a higher conservation of the HA receptor-binding site. Virology 233:224–234
139. Matrosovich M, Tuzikov A, Bovin N, Gambarian A, Klimov A, Cox N, Castrucci M, Donatelli I, Kawaoka Y (2000) Alterations of receptor-binding properties of H1, H2 and H3 avian influenza virus hemagglutinins upon introduction into mammals. J Virol 74:8502–8512
140. Klenk HD, Rott R, Orlich M, Blodorn J (1975) Activation of influenza A viruses by trypsin treatment. Virology 68:426–439
141. Lazarowitz SG, Choppin PW (1975) Enhancement of the infectivity of influenza A and B viruses by proteolytic cleavage of the hemagglutinin polypeptide. Virology 68:440–454
142. Bosch FX, Garten W, Klenk HD, Rott R (1981) Proteolytic cleavage of influenza virus hemagglutinins: primary structure of the connecting peptide between HA1 and HA2 determines proteolytic cleavability and pathogenicity of avian influenza viruses. Virology 113:725–735
143. Garten W, Bosch FX, Linder D, Rott R, Klenk HD (1981) Proteolytic activation of the influenza virus hemagglutinin: the structure of the cleavage site and the enzymes involved in cleavage. Virology 115:361–374

144. Garten W, Klenk HD (1983) Characterization of the carboxypeptidase involved in the proteolytic cleavage of the influenza hemagglutinin. J Gen Virol 64(Pt 10):2127–2137
145. Garten W, Linder D, Rott R, Klenk HD (1982) The cleavage site of the hemagglutinin of fowl plague virus. Virology 122:186–190
146. Garten W, Klenk HD (2008) Cleavage activation of the influenza virus hemagglutinin and its role in pathogenesis. In: Klenk HD, Matrosovich MN, Stech J (eds) Avian influenza, vol 27. Karger, Basel, pp 156–167
147. Klenk HD, Garten W (1994) Host cell proteases controlling virus pathogenicity. Trends Microbiol 2:39–43
148. Steinhauer DA (1999) Role of hemagglutinin cleavage for the pathogenicity of influenza virus. Virology 258:1–20
149. Guenther I, Glatthaar B, Doller G, Garten W (1993) A H1 hemagglutinin of a human influenza A virus with a carbohydrate-modulated receptor binding site and an unusual cleavage site. Virus Res 27:147–160
150. Kawaoka Y, Yamnikova S, Chambers TM, Lvov DK, Webster RG (1990) Molecular characterization of a new hemagglutinin, subtype H14, of influenza A virus. Virology 179:759–767
151. Boettcher E, Matrosovich T, Beyerle M, Klenk HD, Garten W, Matrosovich M (2006) Proteolytic activation of influenza viruses by serine proteases TMPRSS2 and HAT from human airway epithelium. J Virol 80:9896–9898
152. Tashiro M, Ciborowski P, Klenk HD, Pulverer G, Rott R (1987) Role of Staphylococcus protease in the development of influenza pneumonia. Nature 325:536–537
153. Stieneke-Groeber A, Vey M, Angliker H, Shaw E, Thomas G, Roberts C, Klenk HD, Garten W (1992) Influenza virus hemagglutinin with multibasic cleavage site is activated by furin, a subtilisin-like endoprotease. Embo J 11:2407–2414
154. Seidah NG, Chretien M (1999) Proprotein and prohormone convertases: a family of subtilases generating diverse bioactive polypeptides. Brain Res 848:45–62
155. Hallenberger S, Bosch V, Angliker H, Shaw E, Klenk HD, Garten W (1992) Inhibition of furin-mediated cleavage activation of HIV-1 glycoprotein gp160. Nature 360:358–361
156. Horimoto T, Nakayama K, Smeekens SP, Kawaoka Y (1994) Proprotein-processing endoproteases PC6 and furin both activate hemagglutinin of virulent avian influenza viruses. J Virol 68:6074–6078
157. Neumann G, Kawaoka Y (2006) Host range restriction and pathogenicity in the context of influenza pandemic. Emerg Infect Dis 12:881–886
158. Kawaoka Y, Webster RG (1989) Interplay between carbohydrate in the stalk and the length of the connecting peptide determines the cleavability of influenza virus hemagglutinin. J Virol 63:3296–3300
159. Kawaoka Y, Naeve CW, Webster RG (1984) Is virulence of H5N2 influenza viruses in chickens associated with loss of carbohydrate from the hemagglutinin? Virology 139: 303–316
160. Li SQ, Orlich M, Rott R (1990) Generation of seal influenza virus variants pathogenic for chickens, because of hemagglutinin cleavage site changes. J Virol 64:3297–3303
161. Morsy J, Garten W, Rott R (1994) Activation of an influenza virus A/turkey/Oregon/71 HA insertion variant by the subtilisin-like endoprotease furin. Virology 202:988–991
162. Garcia M, Crawford JM, Latimer JW, Rivera-Cruz E, Perdue ML (1996) Heterogeneity in the hemagglutinin gene and emergence of the highly pathogenic phenotype among recent H5N2 avian influenza viruses from Mexico. J Gen Virol 77(Pt 7):1493–1504
163. Perdue ML, Garcia M, Senne D, Fraire M (1997) Virulence-associated sequence duplication at the hemagglutinin cleavage site of avian influenza viruses. Virus Res 49:173–186
164. Banks J, Speidel ES, Moore E, Plowright L, Piccirillo A, Capua I, Cordioli P, Fioretti A, Alexander DJ (2001) Changes in the hemagglutinin and the neuraminidase genes prior to the emergence of highly pathogenic H7N1 avian influenza viruses in Italy. Arch Virol 146:963–973

165. Varghese JN, Laver WG, Colman PM (1983) Structure of the influenza virus glycoprotein antigen neuraminidase at 2.9 A resolution. Nature 303:35–40
166. Blok J, Air GM (1982) Block deletions in the neuraminidase genes from some influenza A viruses of the N1 subtype. Virology 118:229–234
167. Colman PM, Varghese JN, Laver WG (1983) Structure of the catalytic and antigenic sites in influenza virus neuraminidase. Nature 303:41–44
168. Mitnaul LJ, Matrosovich MN, Castrucci MR, Tuzikov AB, Bovin NV, Kobasa D, Kawaoka Y (2000) Balanced hemagglutinin and neuraminidase activities are critical for efficient replication of influenza A virus. J Virol 74:6015–6020
169. Wagner R, Wolff T, Herwig A, Pleschka S, Klenk HD (2000) Interdependence of hemagglutinin glycosylation and neuraminidase as regulators of influenza virus growth: a study by reverse genetics. J Virol 74:6316–6323
170. Hausmann J, Kretzschmar E, Garten W, Klenk HD (1995) N1 neuraminidase of influenza virus A/FPV/Rostock/34 has haemadsorbing activity. J Gen Virol 76(Pt 7):1719–1728
171. Kobasa D, Rodgers ME, Wells K, Kawaoka Y (1997) Neuraminidase hemadsorption activity, conserved in avian influenza A viruses, does not influence viral replication in ducks. J Virol 71:6706–6713
172. Laver WG, Colman PM, Webster RG, Hinshaw VS, Air GM (1984) Influenza virus neuraminidase with hemagglutinin activity. Virology 137:314–323
173. Nuss JM, Air GM (1991) Transfer of the hemagglutinin activity of influenza virus neuraminidase subtype N9 into an N2 neuraminidase background. Virology 183:496–504
174. Varghese JN, Colman PM, van Donkelaar A, Blick TJ, Sahasrabudhe A, McKimm-Breschkin JL (1997) Structural evidence for a second sialic acid binding site in avian influenza virus neuraminidases. Proc Natl Acad Sci USA 94:11808–11812
175. Uhlendorff J, Matrosovich T, Klenk HD, Matrosovich M (2009) Functional significance of the hemadsorption activity of influenza virus neuraminidase and its alteration in pandemic viruses. Arch Virol 154:945–957
176. Matrosovich MN, Matrosovich TY, Gray T, Roberts NA, Klenk HD (2004) Neuraminidase is important for the initiation of influenza virus infection in human airway epithelium. J Virol 78:12665–12667
177. Palese P, Tobita K, Ueda M, Compans RW (1974) Characterization of temperature sensitive influenza virus mutants defective in neuraminidase. Virology 61:397–410
178. Herrler G, Rott R, Klenk HD, Muller HP, Shukla AK, Schauer R (1985) The receptor-destroying enzyme of influenza C virus is neuraminate-O-acetylesterase. EMBO J 4:1503–1506
179. Rosenthal PB, Zhang X, Formanowski F, Fitz W, Wong CH, Meier-Ewert H, Skehel JJ, Wiley DC (1998) Structure of the hemagglutinin-esterase-fusion glycoprotein of influenza C virus. Nature 396:92–96
180. Hay AJ, Collins PJ, Russell RJ (2008) Antivirals and resistance. In: Klenk HD, Matrosovich MN, Stech J (eds) Avian influenza. Karger, Basel, pp 252–271
181. Haller O, Kochs G, Staeheli P (2008) Influenza a virus virulence and innate immunity: Recent insights from new mouse models. In: Klenk HD, Matrosovich MN, Stech J (eds) Avian influenza. Karger, Basel, pp 195–209
182. Neumann G, Noda T, Kawaoka Y (2009) Emergence and pandemic potential of swine-origin H1N1 influenza virus. Nature 459:931–939

Investigating the Interaction Between Influenza and Sialic Acid: Making and Breaking the Link

John M. Nicholls, Jimmy Lai, and Jean-Michel Garcia

Abstract Since the early 1940s sialic acid (Sia) has been regarded as the primary receptor for influenza virus. This Sia is usually bound to an adjacent galactose (Gal) in an α2-3 or α2-6 configuration. This led to a concept about an interspecies barrier as avian viruses preferentially bind to Sia α2-3 linked to Gal, whereas human viruses have a preference for the Sia α2-6 linked to Gal and that transmission from one species to another would preferentially occur only in a host species in which both types of Sia were present . The viral haemagglutinin binds to Sia to facilitate cellular entry. To release progeny viral particles the second main component of the influenza viral envelope – neuraminidase, cleaves Sia. The viral-receptor interaction was initially investigated using agglutination of red blood cells and later using lectin histochemistry. Recent techniques investigating the HA-Sia/NA-Sia link have employed the use of glycan arrays and virus-like pseudoparticles with STD-NMR.

1 Influenza Virus and Haemagglutination

In 1941, two publications by George Hirst and McClelland and Hare demonstrated that influenza virus was able to agglutinate red blood cells (reviewed in [1]). Before 1941 the diagnosis of influenza was made by virus neutralization and complement fixation but the in vitro haemagglutination showed that the quantity of virus as well as the presence of antibody could be used as a simple, rapid and less laborious test for influenza virus infection as the haemagglutination reaction could be detected by the naked eyes in minutes. Studies in subsequent years further characterized the

J.M. Nicholls (✉)
Department of Pathology, The University of Hong Kong, Pok Fu Lam, Hong Kong
e-mail: nicholls@pathology.hku.hk

J. Lai • J.-M. Garcia
HKU-Pasteur Research Centre, The University of Hong Kong, Pok Fu Lam, Hong Kong

M. von Itzstein (ed.), *Influenza Virus Sialidase - A Drug Discovery Target*,
Milestones in Drug Therapy, DOI 10.1007/978-3-7643-8927-7_2,

nature of the haemagglutination – it was temperature dependent but the cell receptors were comparatively heat stable resisting 100°C for 5 min [2, 3]. The haemagglutination reaction was also pH dependent as some strains of influenza did not agglutinate at pH 5.0 though agglutination could continue up to pH = 10 [4–6].

2 The Nature of the Red Blood Cell Receptor

Because of the sensitivity of haemagglutination to temperature and formalin treatments, and its resistance to oxidizing agents, Hirst suggested that the influenza receptors present on red blood cells could be mucoproteins [6]. The finding that agglutination was not blood group dependent (A, B, AB or O) and that treated cells could still agglutinate with appropriate antisera excluded the blood group antigens as the potential link. Gottschalk and Lind in 1949 showed that the ovomucin fraction of egg white contained a haemagglutination inhibitor and further work by Gottschalk on this mucin narrowed this virus inhibiting nature down to a disaccharide – sialic acid bound to galactosamine and that this disaccharide was O-linked [7, 8]. Parallel studies by Burnet and Stone demonstrated that the haemagglutination reaction could be prevented by treatment of red blood cells with enzymes from *Vibrio cholerae* [9]. This so-called receptor destroying enzyme (RDE) was later shown by Gottschalk to be a sialidase or neuraminidase [10]. (Though the terms neuraminidase and sialidase are used interchangeably most researchers tend to use the term sialidase rather than neuraminidase, but the latter term is commonly used for influenza virus to be consistent with the H and N subtype classification.) The predominant glycoprotein present on red blood cells is glycophorin which contains 28% Sia by weight and accounts for 80% of the total erythrocyte Sia [11]. As Sias have a negative charge in a physiological condition, they mutually repel each other, and in the case of the high molecular weight filamentous glycoprotein mucins that were used by Gottschalk and Klenk, the attached sialic acids give a high viscosity to mucus to act as a lubricant. This negative charge prevents the spontaneous clumping of blood platelets and protects macromolecules, such as intrinsic factor, from being destroyed by proteolytic enzymes [10].

3 Not All Sialic Acid Is the Same

Using red blood cells from different animal species agglutination assays produced conflicting results and this was previously explained by strains of virus, temperature of reaction and species used. For instance, while viruses were able to agglutinate red blood cells from most animal species (man, monkey, guinea pig, dog, ferret, rat, mouse, duck and birds) at both 4°C and 20°C, some viruses would only agglutinate horse, ox, sheep, pig and hamster at 4°C but not 20°C [12]. Later work showed that though the most common Sia found in mammals was *N*-acetylneuraminic acid (Neu5Ac), some animal species including great apes contained *N*-glycolylneuraminic acid (Neu5Gc) [13]. Neu5Gc is absent from humans due to the lack of 92 base pairs segment of the CMP-*N*-acetylneuraminic acid hydroxylase gene responsible for its synthesis and some virus strains preferentially bind to Neu5Ac than Neu5Gc which could explain the previously reported differences in haemagglutination.

4 The Second Interaction of Influenza with Red Blood Cells: Elution

The initial studies of Hirst on haemagglutination showed that adsorbed virus could be eluted from red blood cells at 37°C without significant loss but that the cells from which the virus had been eluted could not be agglutinated by fresh virus, indicating that the interaction of the virus with the red blood cell had structurally modified them so that the same reaction could not occur. This lack of re-agglutination was proposed to be due to enzymatic destruction of the receptor [2]. Klenk subsequently incubated influenza virus with urinary mucin to crystallize neuraminic acid. Initially, it was thought that the sialidase activity was combined with the haemagglutinin, but in 1961 Howe et al. were able to show that the HA was distinct from the NA and that they were distinct independent entities of the virus particle. This work was followed later by Laver who in 1963 was able to separate neuraminidase from influenza virus (reviewed in [9]). The role of neuraminidase in the viral replication cycle was investigated by Noll et al. who found that no NA was present within the first 4 h of infection, but it increased from 6 h onwards and this correlated with release of virus [14]. The NA was thus considered essential for viral particles being released from the cell.

5 Different Viruses, Different Agglutination and Elution

Though influenza was suspected to be a diverse agent it was not until 1940 that Francis and Magill, working independently were able to demonstrate two new strains of influenza that were antigenically different from strains of influenza A virus (reviewed in [15]). Studies by Paulson and colleagues in the late 1970s and 1980s using ortho- and para-myxoviruses exploited the RDE nature of *V. cholerae* to desialylate red blood cells and then resialylate them using different sialyltransferases (ST) – enzymes that attach Sia to the terminal galactose. They found that depending on the ST used they could change the haemagglutinating ability of the native virus and that even though glycophorin (which had mainly *O*-linked glycosylation) was involved, *N*-linked glycans that composed a minor component of the red cell membrane could play a part. Indeed in their studies using resialylation a relatively small incorporation (10%) of Sia into a specific band of the erythrocyte membrane protein (Band 3) was able to restore agglutination of influenza to normal levels [11]. Furthermore, viruses that were antigenically similar (A/RI/5$^-$/57(H2N2) and A/RI/5$^+$/57(H2N2)) could have different binding properties. These studies demonstrated that it was not just the Sia that was involved in the virus binding and haemagglutination, but the link of the Sia to the adjacent galactose that was important. Furthermore, depending on the receptor binding properties, the viral neuraminidase would have different eluting effects, thus suggesting that the link between the Sia and galactose would influence the effectiveness of the viral NA. Jameson and Levine in 1953 also demonstrated that certain strains of influenza were unable to hydrolyse porcine submaxillary mucin which contains 90% Neu5Gc suggesting that the viral neuraminidase could distinguish between Neu5Ac and Neu5Gc [13].

6 The Addition of Sialic Acids to Galactose Requires Different Sialyltransferases

As indicated previously, Sia are added to the terminal sugar of glycoproteins and glycolipids by enzymes called sialyltransferases (ST) and normally once Sia are added they cannot be further extended [16]. The three common sugar residues that can have Sias linked to them are galactose (Gal), *N*-acetylglucosamine (GlcNAc), *N*-acetylgalactosamine (GalNAc) or another Sia. When Sia is added to a Gal it can either be bound to the hydroxyl group attached to carbon-3 (3-OH) of Gal to form an α2-3 glycosidic linkage, or to the 6-OH to form an α2-6 glycosidic linkage. When Sia is added to another Sia it is usually in a α2-8 linkage to form dimers, oligomers or longer homopolymers of Sia. The synthesis of the Sia terminated glycans is carried out by a family of 20 ST that are highly conserved from mouse to man [17]. For example, those that transfer Sia in an α2-6 linkage to Gal are ST6Gal transferases and those that transfer Sia in an α2-3 linkage are ST3Gal transferases. In humans, there are two ST6Gal transferases (ST6Gal I and II) both of which act on type II glycan chains (Galß1-4GlcNAc-R). The ST3Gal family is more complex with six members (ST3Gal I-VI). ST3Gal III, IV and VI sialylate type II glycan chains (Galß1-4GlcNAc-R), and ST3Gal I and ST3Gal II have a preference for type III glycan chains present on *O*-glycans (Galß1-3GalNAc-R), ST3Gal V acts on lactosyl-ceramide glycolipids (Galß1-4GlcNAc-Cer). Thus even though the STs use a common donor, CMP-Neu5Ac, the ST within the same family can differentiate between disaccharide acceptors with a ß1-3 link and a ß1-4 linkage and this has recently been attributed to single amino acid changes in the ST [17]. In *O*-linked glycans, the 2 main 6-linked STs transfer Sia to a GalNAc rather than Gal. The expression of ST is dynamic and can be upregulated by disease states in tumour cells, inflammation, following steroid treatment and tumour necrosis factor α stimulation [18–21].

7 Binding Affinity for Sia Can Be Affected by Many Components of the Glycan

When Rogers and Paulson desialylated human red blood cells and re-sialylated them they took advantage of the different affinities of the ST and thus using ST6Gal, ST6GalNAc and ST3Gal were able to create red blood cells with different Sia linkages. They found that human and swine influenza viruses preferentially agglutinated red blood cells with a terminal Siaα2-6Gal and avian viruses preferentially agglutinated Siaα2-3Gal expressing cells [22]. While there was preference for a α2-3 or α2-6 binding in most of the viruses, in some there was still agglutination of the other receptor suggesting that the binding preference may not be absolute. Furthermore, within the Siaα2-3 binding group, the inner configuration of the oligosaccharide also affected binding – all the viruses that agglutinated red blood cells with Siaa2-3Galß1-3**Gal**NAc would agglutinate when the GalNAc was changed to a **Glc**NAc. However, the converse was not universally true as only half (two out of four) of the viruses that bound Siaa2-3Galß1-**4/3Glc**NAc would tolerate a change to Siaa2-3Galß1-**3Gal**NAc. In particular, an H1N1 virus (A/PR8/34) could tolerate a change from a Siaa2-3ß1-**3Gal**NAc to Siaa2-3ß1-**4Glc**NAc but H3N2 viruses lost binding when GlcNAc was changed to GalNAc – indicating that the viruses had different affinities for additional glycan structural features – not just the linkage position of the terminal Sia [23].

The importance of the subterminal sugars has been explored more using Sia linked polymers and glycan arrays developed by the Consortium for Functional Glycomics that allow for a large range of glycans to be investigated [24]. In the case of avian H5N1 viruses, Gambaryan and colleagues have found that chicken and duck H5N1 viruses differ in their affinity for the subterminal and subsequent residues of Siaα2-3 oligosaccharides – duck viruses bound type I glycan chains (Galß1-**3Glc**NAc) and type III glycan chains (Galß1-**3Gal**NAc), but chicken viruses preferred a type II core (Galß1-**4Glc**NAc), especially when it was sulfated on the 6 position of the GlcNAc [25]. This sulfation is important as sulfotransferases responsible for the generation of such high affinity sulfated sialyloligosaccharide receptor are expressed in human airway cells [26]. Fucosylation at the 3 position of the GlcNAc to generate the $SiaLe^{X}$ structure (NeuAcα2-3Galß1-4(Fucα1-3) GlcNAc) also allowed preferential binding of some gull and H7N1 viruses [23]. These changes in the subterminal residues appear to have more effect on Siaα2-6 than Siaα2-3 glycans owing to their postulated binding interactions with haemagglutinin. The Siaα2-3 configuration is more rigid and appears to bind in the same way with less concern for the underlying HA configuration [27]. Siaα2-6Gal configurations, however, result in different relationships with the underlying

HA and the affinity varies quite considerably depending on the amino acid structure of the HA [28]. A potential explanation of these variations in HA specificity could be due to changes in the glycosylation sites of the HA molecule itself [29], resulting in changes to protein folding and conformation in the haemagglutinin or possible masking of the receptor binding site, and this glycosylation change has been demonstrated by Klenk in 2002 [30] and Zhang in 2004 [31].

8 Lectins Also Act as Agglutinating Agents

Since the affinity of viruses for a cell may be dependent on the Sia–Gal linkage it is natural that methods should be sought to determine the status of this linkage in the respiratory tract. Viruses have been previously shown to be agglutinating agents and lectins have also demonstrated the same function. Lectins are proteins or glycoproteins that bind to defined carbohydrate structures [32], and these can be of animal, microbial or plant origins [33]. As lectins bind sugars typically found on cell surfaces, they can agglutinate cells to which they interact (Fig. 1). It was the

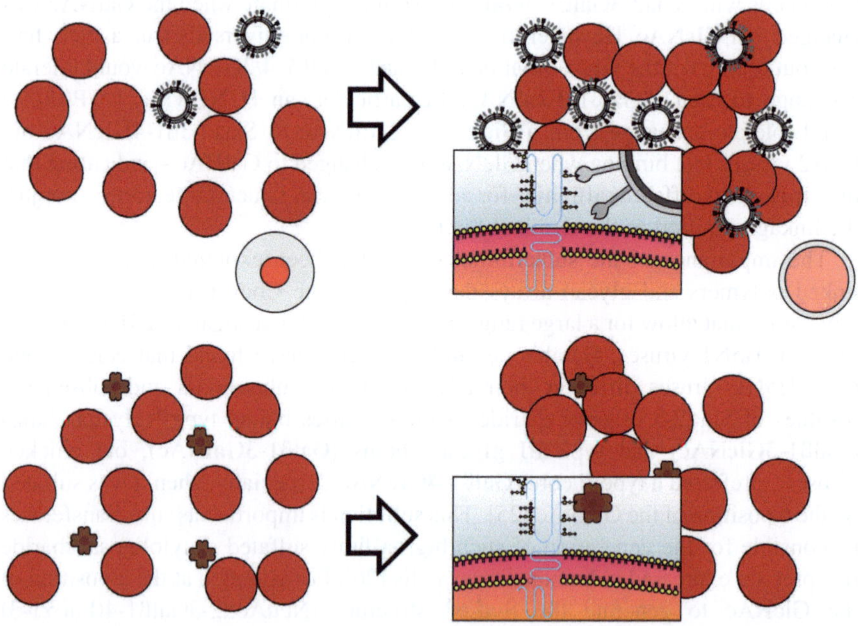

Fig. 1 Hemagglutination due to influenza virus or lectins is due to similar mechanisms. (**a**) The HA of influenza virus recognizes sialic acid terminated oligosaccharides on the surface of red blood cells (*red circles*) leading to agglutination with clumping of the red blood cells. (**b**) Lectins (in this case *Maackia amurensis* brown cross) also recognize the sialic acids present in the oligosaccharides and this leads to agglutination. In the inset, red blood cell glycophorin containing Siaα2-3Gal terminated oligosaccharides is shown

agglutinating activity of these substances that led William Clouser Boyd to use the term lectins for their ability to pick out or choose (Latin *legere*) individual blood group specificity affecting red blood cells of some species of an individual and not others of the same species. For instance, lima bean would agglutinate blood group A but not blood group O or B. In 1961, lectins were identified in seeds from the tree *Maackia amurensis* [34]. Further studies identified two types of *M. amurensis* lectin – one which was haemagglutinating (MAH) and one which was mitogenic (MAM) [35]. The mitogenic MAM was subsequently designated as the *M. amurensis* leukoagglutinin (MAL) due to its ability to agglutinate a mouse lymphoma cell line [36]. MAL was further reported to be specific for Siaα2-3Gal [37]. A lectin from the elderberry plant – *Sambucus nigra* agglutinin (SNA) – was also identified with specificity for the Siaα2-6Gal linkage [38]. Since 1983 the use of *M. amurensis* agglutinin (MAA) and SNA has been used by researchers to determine whether cells express α2-3 or α2-6 linked Sia on the surface. Glycan array data performed by the Consortium for Functional Glycomics has shed more light on the specificity of these lectins for α2-3 and α2-6 linked Sia – the highest binding affinity for MAL-II is for the *O*-linked disialylated Thomsen–Friedenreich (TF) antigen while MAL-I binds to Sia with GlcNAc linked to Gal but also to non-Sia terminated glycans.

9 Probing the Glycan Distribution Using Mass Spectrometry

An additional tool for determining the glycans present in the respiratory tract is mass spectrometry. In this methodology, glycoproteins and glycolipids are extracted from the respiratory mucosa and then profiled according to mass and known biochemical profiles. Sialidase digestion using α2-3 only or α2-3/2-6 specific sialidases then allows an analysis of the types of linkages present (Fig. 2). The advantage of this technique is that it provides a systematic analysis of the total glycans present and thus a better indication of the potential binding sites than lectin histochemistry. However, as respiratory tissues are a heterogenous population and analysis needs to be correlated with lectin histochemistry findings for better determination of which cells contain which types of Sia-linked glycans.

10 Changes in the HA Receptor Binding Site Affects Influenza Binding

The differential recognition of Siaα2-3Gal or Siaα2-6Gal by influenza viruses of different species was attributed to changes in the amino acid sequence of the haemagglutinin (HA). A change in the amino acid 226 of HA1 protein (produced by cleavage of HA0 into HA1 and HA2) from leucine to glutamine (or rarely methionine) in a H3 influenza virus led to a change in binding preference from

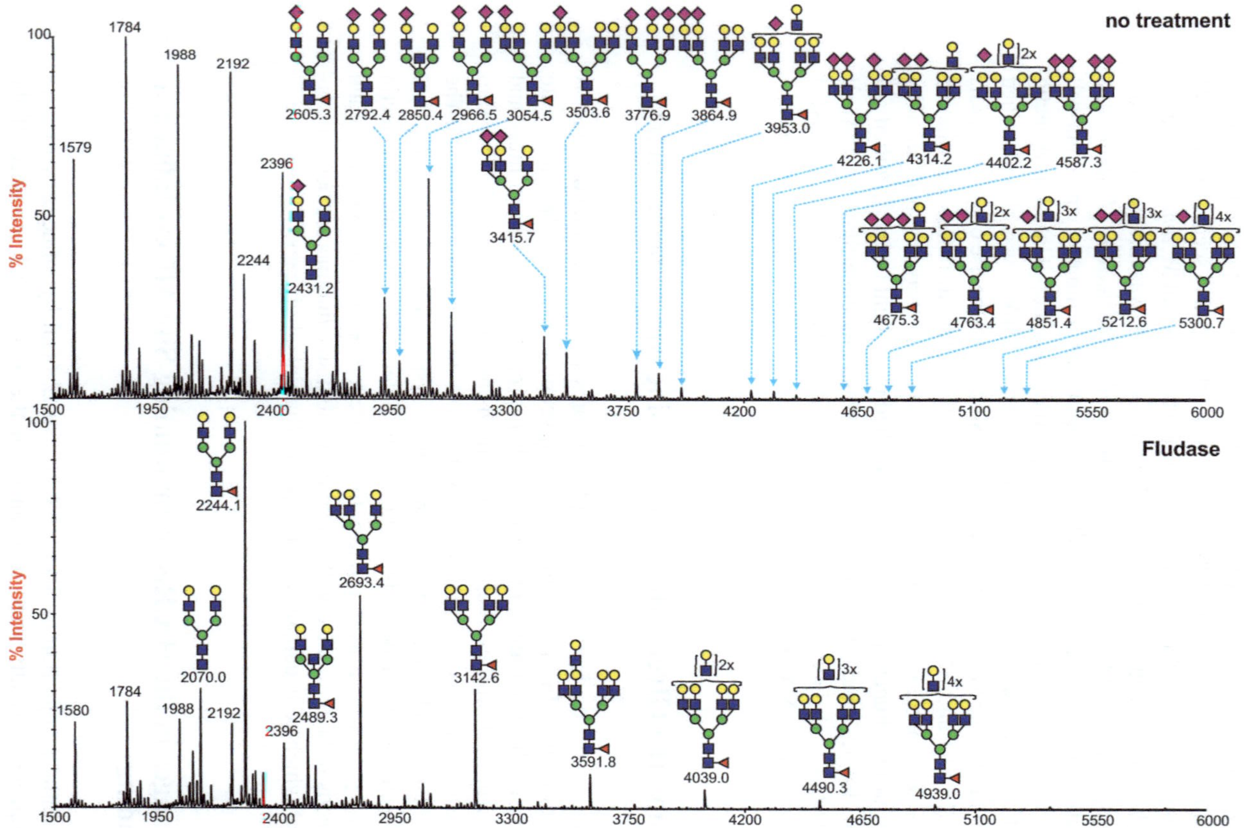

Fig. 2 MALDI-TOF mass spectra of permethylated N-glycans from human ex vivo lung biopsy material before (A) and after (B) neuraminidase (DAS181) treatment. N-glycans were released from the glycoproteins and glycopeptides by peptide:N-glycosidase F digestion and purified. The N-glycans were permethylated (methyl iodide-dimethyl sulfoxide-NaOH procedure) and purified by C_{18} Sep-Pak chromatography. The samples were analysed by matrix-assisted laser desorption ionization-time of flight (MALDI-TOF) MS to obtain molecular ion profiles of the glycans, and selected ions were further characterized by tandem MS (MS/MS) analysis [56]

Siaα2-6Gal to Siaα2-3Gal [39]. Conversely, A/Udorn/307/H2 (H3N2) did not replicate in a duck intestine, while an avian reassortment virus A/Mal/NY/6750/78 (H2N2), with seven avian virus genes and a human HA gene with a mutation at 226 and 228, did [40]. Based on these findings, it was proposed that Gln226 correlated with Siaα2-3Gal specificity in avian viruses and Leu226 with Siaα2-6Gal in human viruses of H2 [41] and H3 [39, 42] subtypes. However, these findings do not apply to the H1 [43] or H5 subtype, and human H2 and the avian H6 and H9 bound Siaα2-6 motifs even though they contained Gln226. A change from Glu190 to Asp has been considered important in changing from Siaα2-3 to Siaα2-6 binding, which did not appear in human H3 isolates until 1992, when they lost the ability to agglutinate chicken red blood cells but were still able to agglutinate human red blood cells. This presence of Glu190 may have been an artefact of passage of isolates in chicken eggs, but recent isolates do not show acquisition of Siaα2-3 binding or the Asp190Glu change when adapted to growth in eggs [28] and have also shown differences in affinity for 2–6 linked glycans from strain to strain in the past 8 years [44].

11 NMR: A Physical Approach to a Biological Situation

All the above mentioned data clearly demonstrate that our understanding on influenza virus interaction with its cellular receptor is much more complex than previously thought, and the limitations of determining the presence of α2-6 versus α2-3 linked glycans by lectin histochemistry needs to be readdressed. Among the new avenues of investigation, a recent technology based on physical measurement – nuclear magnetic resonance (NMR), has enabled a more precise interaction of the HA with the sialylated ligands down to the atomic level.

In 1938, Rabi observed for the first time that when a magnetic nucleus, such as hydrogen ^{1}H, carbon ^{13}C, nitrogen ^{15}N, was placed in a magnetic field and an electromagnetic pulse was applied, it caused the nucleus to absorb energy from the pulse and radiate this energy back out at a specific resonance frequency which depended on the environment of the atom [45]. Ever since, this phenomenon of NMR was used on routine basis for the determination of chemical structure of molecules. For example, as each ^{1}H in a molecule is in a different environment, each of them will have its own "signature." Perturbation of this environment, for example through interactions with another molecule, could be detected in strictly controlled conditions offering a new way of looking at ligand–protein interactions.

Although several NMR protocols have been created to that purpose, one extremely powerful method, called Saturation Transfer Difference NMR (STD-NMR), although known for many years, was developed by Mayer and Meyer [46] for the characterization of ligand binding to large proteins and its application for the screening of libraries of compounds.

As illustrated in Fig. 3, the principle of this method is relatively simple. In macromolecules such as proteins, all the protons are interrelated to each other

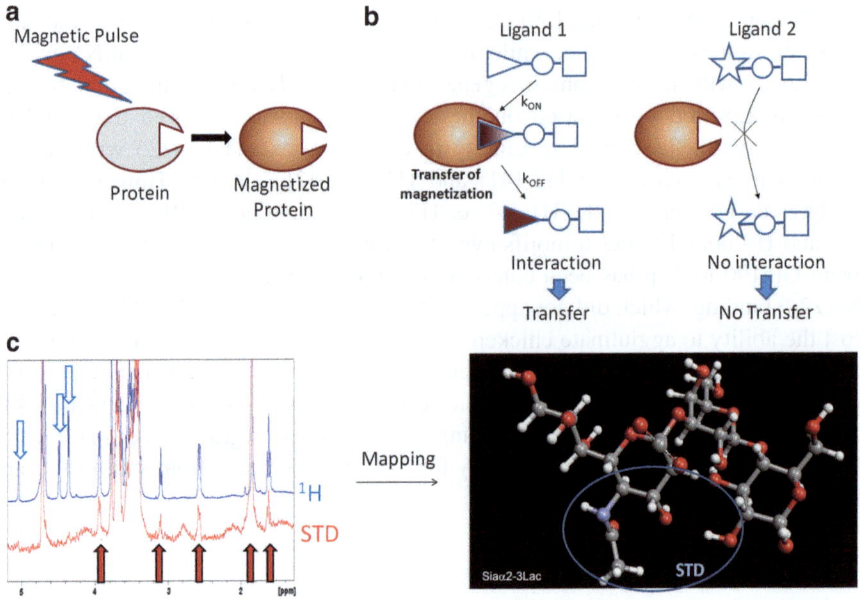

Fig. 3 Principle of interaction mapping by STD-NMR. (**a**) The protein of interest placed in a magnetic field is magnetized. (**b**) If a ligand interacts with that protein, a part of magnetization is transferred to through the interaction to the chemical group in the ligand proportionally (given the proper condition) to the strength and distance to the interaction binding site. (**c**) By comparison between the regular 1H proton spectra (that display all the different type of proton of the molecule) and the STD spectra, one can map the binding of the ligand with the protein

through a matrix of so-called dipole–dipole interactions. Therefore, if a pulse corresponding to one of the protein resonance is applied the magnetization will rapidly spread over the entire protein. As the range of resonance of the macromolecule (receptor) is much wider than of small molecules (ligands) the later one will be unaffected as long as it is in solution. Once the ligand interacts with its receptor, intermolecular transfer of magnetization occurs to the ligand with a graduation: the closer to the interaction the stronger the magnetization. This led these authors to introduce the notion of group epitope mapping [47] and can now provide access on the interaction between a substrate and its receptor with atomic resolution.

Although initially STD-NMR experiments were applied to purified proteins [48, 49] recently full viruses were used when looking at surface proteins. These guaranteed the presentation of the proteins in their native conformation. Nevertheless, this approach is impractical with pathogenic viruses for biosafety reasons. To overcome this problem while maintaining the advantages of the technique, virus-like particles (VLP) were developed, for example, for highly pathogenic avian influenza H5N1 [50].

VLPs are produced by co-transfection of plasmids coding for the core of the virion as one component, and for the surface glycoproteins that will be incorporated on the virion surface during the budding as other components. As the viral surface glycoproteins are presented in a similar way to native virions, such chimeric pseudoviral particles can be used as mimics of virus for investigations changing the properties of the envelop proteins [51]. For example, Garcia et al. showed it could be used, after introduction of a reporter-gene, for the sero-diagnostic or sero-epidemiological investigation of outbreak of H5N1 in BSL-2 conditions instead of

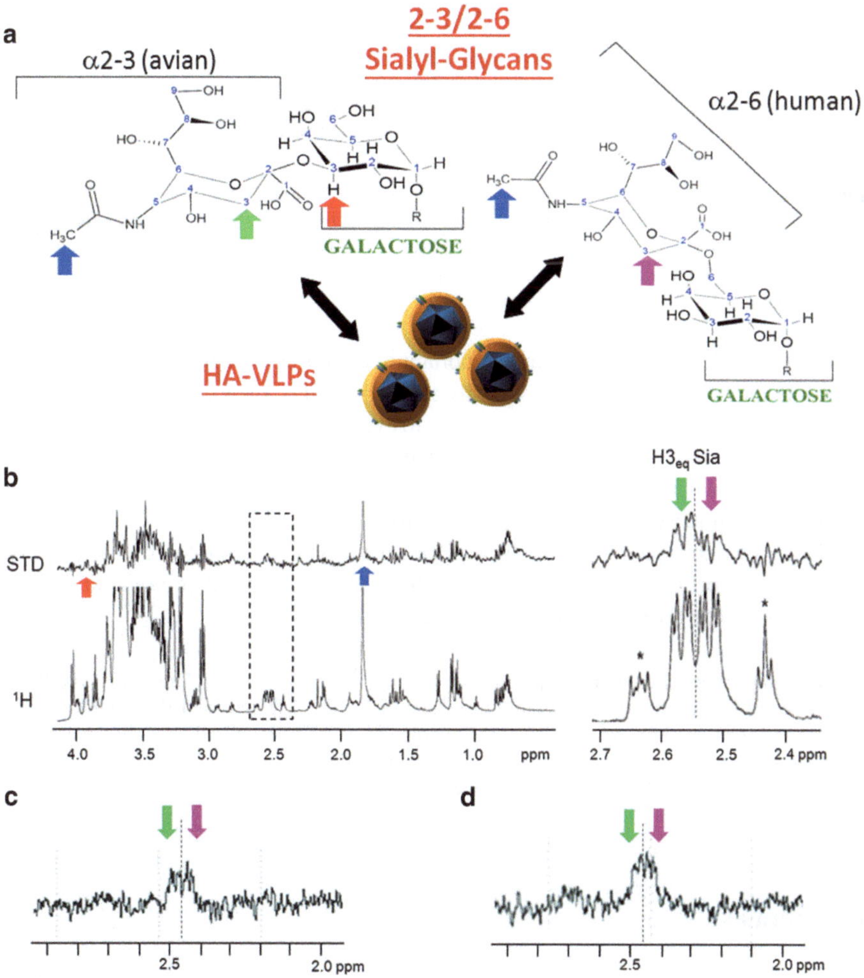

Fig. 4 Application of STD-NMR to the study of influenza receptor. (**a**) Virus-like particles bearing HA proteins (HA-VLP) were analysed by STD-NMR with a mix of Siaα2-3- or Siaα-2-6-sialyllactose (SL). These two ligands differ by the branching of second carbon of sialic acid to next sugar as reflected, among other signals, for example by the $H3_{eq}$ between 2.5 and 2.6 ppm. As expected, avian H5-VLP preferential bind for α2-3SL (**b**). We could also confirm that human seasonal H3 binds equivalently to both α2,3/2,6-SL (c) as do avian H5-VLP with double mutations Q192R/S223N in HA as described by Yamada et al. [57]

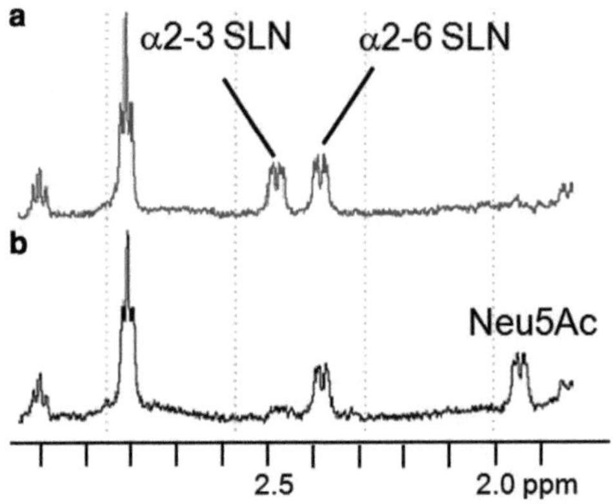

Fig. 5 Application of NMR to the study of neuraminidase sialidase activity. (**a**) Regular proton spectrum of an equimolar mixture of α2-3- or α2-6-sialyllactosamine (SLN). (**b**) After addition of virus-like particles bearing the N1 neuraminidase (N1-VLP) from avian H5N1 virus to the mixture, there is cleavage of sialic acid and the apparition of the end product Neu5Ac with a clear preference for the α2-3SLN substrate

BSL-3 [52, 53]. This is particularly important in countries or region where such high biosafety facility is not available or easily accessible. This assay could also be used in more fundamental research as, for example, to evaluate the cross-subtype neutralizing response [54]. VLP's applications are not restricted to the vial haemagglutinin as previously shown in (Fig. 4). NA-VLPs have used also to monitor the catalytic activity of the neuraminidase as illustrated by Fig. 5 for N1 from H5N1 [55].

These few examples illustrate how physical analytical technologies such as MS or NMR can bring powerful insights on the function of viral proteins and their interactions with host with an increased resolution.

Acknowledgements Research supported by funding from Wellcome Trust, UK (082098, 081184/Z/06/Z), Research Grants Committee (GRF 4774109), HKSAR and Research Fund for Control of Infectious Disease (08070842), Government of HKSAR, NIAID. Drs Stuart Haslam and Rositsa Karamanska, Imperial College, United Kingdom are thanked for valuable help on glycan structures and glycobiology experience.

References

1. Schauer R (2000) Achievements and challenges of sialic acid research. Glycoconj J 17:485–499
2. Hirst GK (1942) Adsorption of influenza hemagglutinins and virus by red blood cells. J Exp Med 76:195–209
3. Archetti I, Donald HB, Isaacs A, Valentine RC (1955) The effects of metaperiodate and heated influenza virus on influenza virus growth. J Gen Microbiol 13:330–337
4. Miller GL, Stanley WM (1944) Quantitative aspects of the red blood cell agglutination test for influenza virus. J Exp Med 79:185–195

5. Hirst GK (1948) The nature of the virus receptors of red cells; the effect of partial heat inactivation of influenza virus on the destruction of red cell receptors and the use of inactivated virus in the measurement of serum inhibitor. J Exp Med 87:315–328
6. Hirst GK (1948) The nature of the virus receptors of red cells; evidence on the chemical nature of the virus receptors of red cells and of the existence of a closely analogous substance in normal serum. J Exp Med 87:301–314
7. Graham ER, Gottschalk A (1960) Studies on mucoproteins. I. The structure of the prosthetic group of ovine submaxillary gland mucoprotein. Biochim Biophys Acta 38:513–524
8. Gottschalk A, Fezekas De St Groth S (1960) On the relationship between the indicator profile and prosthetic group of mucoproteins inhibitory for influenza virus haemagglutinin. J Gen Microbiol 22:690–697
9. Hoyle L (1968) Virol Monogr 4:1–375
10. Schauer R (1973) Chemistry and biology of the acylneuraminic acids. Angew Chem Int Ed Engl 12:127–138
11. Paulson JC, Sadler JE, Hill RL (1979) Restoration of specific myxovirus receptors to asialoerythrocytes by incorporation of sialic acid with pure sialyltransferases. J Biol Chem 254: 2120–2124
12. Chu CM (1948) Inactivation of haemagglutinin and infectivity of influenza and Newcastle disease viruses by heat and by formalin. J Hyg (Lond) 46:247–251
13. Jameson P, Levine AS (1965) Substrate specificity of neurotropic influenza virus neuraminidases. J Bacteriol 90:563–564
14. Noll H, Aoyagi T, Orlando J (1962) The structural relationship of sialidase to the influenza virus surface. Virology 18:154–157
15. Horsfall FL (1941) Recent studies in influenza. Am J Public Health Nations Health 31:1275–1280
16. Harduin-Lepers A, Mollicone R, Delannoy P, Oriol R (2005) The animal sialyltransferases and sialyltransferase-related genes: a phylogenetic approach. Glycobiology 15:805–817
17. Paulson JC, Rademacher C (2009) Glycan terminator. Nat Struct Mol Biol 16:1121–1122
18. Lise M, Belluco C, Perera SP, Patel R, Thomas P, Ganguly A (2000) Clinical correlations of alpha2,6-sialyltransferase expression in colorectal cancer patients. Hybridoma 19:281–286
19. Yasukawa Z, Sato C, Kitajima K (2005) Inflammation-dependent changes in alpha2,3-, alpha2,6-, and alpha2,8-sialic acid glycotopes on serum glycoproteins in mice. Glycobiology 15:827–837
20. Delmotte P, Degroote S, Merten MD, Van Seuningen I, Bernigaud A, Figarella C, Roussel P, Perini JM (2001) Influence of TNFalpha on the sialylation of mucins produced by a transformed cell line MM-39 derived from human tracheal gland cells. Glycoconj J 18:487–497
21. Ishibashi Y, Inouye Y, Okano T, Taniguchi A (2005) Regulation of sialyl-Lewis x epitope expression by TNF-alpha and EGF in an airway carcinoma cell line. Glycoconj J 22:53–62
22. Rogers GN, Paulson JC (1983) Receptor determinants of human and animal influenza virus isolates: differences in receptor specificity of the H3 hemagglutinin based on species of origin. Virology 127:361–373
23. Gambaryan A, Yamnikova S, Lvov D, Tuzikov A, Chinarev A, Pazynina G, Webster R, Matrosovich M, Bovin N (2005) Receptor specificity of influenza viruses from birds and mammals: new data on involvement of the inner fragments of the carbohydrate chain. Virology 334:276–283
24. Stevens J, Blixt O, Tumpey TM, Taubenberger JK, Paulson JC, Wilson IA (2006) Structure and receptor specificity of the hemagglutinin from an H5N1 influenza virus. Science 312:404–410
25. Gambaryan AS, Tuzikov AB, Pazynina GV, Webster RG, Matrosovich MN, Bovin NV (2004) H5N1 chicken influenza viruses display a high binding affinity for Neu5Acalpha2-3Galbeta1-4(6-HSO3)GlcNAc-containing receptors. Virology 326:310–316

26. Lamblin G, Degroote S, Perini JM, Delmotte P, Scharfman A, Davril M, Lo-Guidice JM, Houdret N, Dumur V, Klein A et al (2001) Human airway mucin glycosylation: a combinatory of carbohydrate determinants which vary in cystic fibrosis. Glycoconj J 18:661–684
27. Russell RJ, Stevens DJ, Haire LF, Gamblin SJ, Skehel JJ (2006) Avian and human receptor binding by hemagglutinins of influenza A viruses. Glycoconj J 23:85–92
28. Kumari K, Gulati S, Smith DF, Gulati U, Cummings RD, Air GM (2007) Receptor binding specificity of recent human H3N2 influenza viruses. Virol J 4:42
29. Vigerust DJ, Shepherd VL (2007) Virus glycosylation: role in virulence and immune interactions. Trends Microbiol 15:211–218
30. Klenk HD, Wagner R, Heuer D, Wolff T (2002) Importance of hemagglutinin glycosylation for the biological functions of influenza virus. Virus Res 82:73–75
31. Zhang M, Gaschen B, Blay W, Foley B, Haigwood N, Kuiken C, Korber B (2004) Tracking global patterns of N-linked glycosylation site variation in highly variable viral glycoproteins: HIV, SIV, and HCV envelopes and influenza hemagglutinin. Glycobiology 14:1229–1246
32. Sharon N (2007) Lectins: carbohydrate-specific reagents and biological recognition molecules. J Biol Chem 282:2753–2764
33. Loris R (2002) Principles of structures of animal and plant lectins. Biochim Biophys Acta 1572:198–208
34. Boyd WC, Waszczenko-Zacharczenko E, Goldwasser SM (1961) List of plants tested for hemagglutinating activity. Transfusion 1:374–382
35. Kawaguchi T, Matsumoto I, Osawa T (1974) Studies on hemagglutinins from Maackia amurensis seeds. J Biol Chem 249:2786–2792
36. Wang WC, Cummings RD (1987) An assay for leukoagglutinating lectins using suspension cultured mouse lymphoma cells (BW5147) stained with neutral red. Anal Biochem 161:80–84
37. Wang WC, Clark GF, Smith DF, Cummings RD (1988) Separation of oligosaccharides containing terminal alpha-linked galactose residues by affinity chromatography on Griffonia simplicifolia I bound to concanavalin A-sepharose. Anal Biochem 175:390–396
38. Shibuya N, Goldstein IJ, Broekaert WF, Nsimba-Lubaki M, Peeters B, Peumans WJ (1987) The elderberry (Sambucus nigra L.) bark lectin recognizes the Neu5Ac(alpha 2–6)Gal/GalNAc sequence. J Biol Chem 262:1596–1601
39. Rogers GN, Paulson JC, Daniels RS, Skehel JJ, Wilson IA, Wiley DC (1983) Single amino acid substitutions in influenza haemagglutinin change receptor binding specificity. Nature 304:76–78
40. Naeve CW, Hinshaw VS, Webster RG (1984) Mutations in the hemagglutinin receptor-binding site can change the biological properties of an influenza virus. J Virol 51:567–569
41. Connor RJ, Kawaoka Y, Webster RG, Paulson JC (1994) Receptor specificity in human, avian, and equine H2 and H3 influenza virus isolates. Virology 205:17–23
42. Vines A, Wells K, Matrosovich M, Castrucci MR, Ito T, Kawaoka Y (1998) The role of influenza A virus hemagglutinin residues 226 and 228 in receptor specificity and host range restriction. J Virol 72:7626–7631
43. Nobusawa E, Aoyama T, Kato H, Suzuki Y, Tateno Y, Nakajima K (1991) Comparison of complete amino acid sequences and receptor-binding properties among 13 serotypes of hemagglutinins of influenza A viruses. Virology 182:475–485
44. Gulati S, Smith DF, Air GM (2009) Deletions of neuraminidase and resistance to oseltamivir may be a consequence of restricted receptor specificity in recent H3N2 influenza viruses. Virol J 6:22
45. Rabi J (1938) A new method of measuring nuclear magnetic moment. Phys Rev 53:318
46. Meyer M, Meyer B (1999) Characterization of ligand binding by saturation transfer difference NMR spectroscopy. Angew Chem Int Ed Engl 38:1784–1788
47. Meyer M, Meyer B (2001) Group epitope mapping by saturation transfer difference NMR to identify segments of a ligand in direct contact with a protein receptor. J Am Chem Soc 123:6108–6117

48. Benie A, Moser R, Bauml E, Blaas D, Peters T (2003) Virus-ligand interactions: identification and characterization of ligand binding by NMR spectroscopy. J Am Chem Soc 125:14–15
49. Rademacher C, Krishna NR, Palcic M, Parra F, Peters T (2008) NMR experiments reveal the molecular basis of receptor recognition by a clicivirus. J Am Chem Soc 130:3669–3675
50. Haselhorst T, Garcia JM, Islam T, Lai JC, Rose FJ, Nicholls JM, Peiris JS, von Itzstein M (2008) Avian influenza H5-containing virus-like particles (VLPs): host-cell receptor specificity by STD NMR spectroscopy. Angew Chem Int Ed Engl 47:1910–1912
51. Nefkens I, Garcia JM, Ling CS, Lagarde N, Nicholls J, Tang DJ, Peiris M, Buchy P, Altmeyer R (2007) Hemagglutinin pseudotyped lentiviral particles: characterization of a new method for avian H5N1 influenza sero-diagnosis. J Clin Virol 39:27–33
52. Garcia JM, Lagarde N, Ma ES, de Jong MD, Peiris JS (2010) Optimization and evaluation of an influenza A (H5) pseudotyped lentiviral particle-based serological assay. J Clin Virol 47:29–33
53. Schultsz C, Nguyen VD, le Hai T, Do QH, Peiris JS, Lim W, Garcia JM, Nguyen DT, Nguyen TH, Huynh HT et al (2009) Prevalence of antibodies against avian influenza A (H5N1) virus among Cullers and poultry workers in Ho Chi Minh City, 2005. PLoS One 4:e7948
54. Garcia JM, Pepin S, Lagarde N, Ma ES, Vogel FR, Chan KH, Chiu SS, Peiris JS (2009) Heterosubtype neutralizing responses to influenza A (H5N1) viruses are mediated by antibodies to virus haemagglutinin. PLoS One 4:e7918
55. Lai JC, Chan WW, Kien F, Nicholls JM, Peiris JS, Garcia JM (2010) Formation of virus-like particles from human cell lines exclusively expressing Influenza neuraminidase. J Gen Virol 91:2322–2330
56. Chan RW, Chan MC, Wong AC, Karamanska R, Dell A, Haslam SM, Sihoe AD, Chui WH, Triana-Baltzer G, Li Q et al (2009) DAS181 inhibits H5N1 influenza virus infection of human lung tissues. Antimicrob Agents Chemother 53:3935–3941
57. Yamada S, Suzuki Y, Suzuki T, Le MQ, Nidom CA, Sakai-Tagawa Y, Muramoto Y, Ito M, Kiso M, Horimoto T et al (2006) Haemagglutinin mutations responsible for the binding of H5N1 influenza A viruses to human-type receptors. Nature 444:378–382

Enzymology of Influenza Virus Sialidase

Jefferson Chan and Andrew J. Bennet

Abstract Influenza virus sialidase plays a key role in the infectious lifecycle of the virus. This chapter provides a discussion of the tools, such as linear free energy relationships and kinetic isotope effects, used in exploring enzyme mechanisms and an introduction to mechanistic aspects, including transition state analysis and whether the intermediate that follows the glycosylation TS, in retaining glycosidases, is an oxacarbenium ion or is covalently linked to the enzyme. A general discussion of microbial sialidase catalysis is provided, as well as an overview of the catalytic mechanism of influenza virus sialidases.

1 Introduction

Sialic acids are a family of nine-carbon α-ketoacid sugars that are found in nature. The most common sialic acid is *N*-acetylneuraminic acid (5-acetamido-3,5-dideoxy-D-*glycero*-D-*galacto*-nonulosonic acid, Neu5Ac **1**, Fig. 1), and in this chapter it is referred to as sialic acid.

This carbohydrate possesses a carboxylate group at C-1, an acetamido-moiety at position 5, and a glycerol chain substituent on C-6 (Structure **1** depicts the β-anomer of sialic acid). Glycosides of sialic acid play important roles in a number of biological events; hence nature has provided a number of enzymes, such as sialidases, that recognize and process this specialized carbohydrate. *exo*-Sialidases (EC 3.2.1.18, sialidases) are a family of enzymes that catalyse hydrolytic cleavage of terminal sialic acid residues from a plethora of glycoconjugates that are predominantly linked to a galactose or *N*-acetylgalactosamine residue via either a 2,3- or 2,6-α-ketosidic linkage or to a sialic acid moiety via a 2,8- or 2,9-α-glycosidic bond

J. Chan • A.J. Bennet (✉)
Department of Chemistry, Simon Fraser University, 8888 University Drive, Burnaby, BC V5A 1S6 Canada
e-mail: bennet@sfu.ca

M. von Itzstein (ed.), *Influenza Virus Sialidase - A Drug Discovery Target*, Milestones in Drug Therapy, DOI 10.1007/978-3-7643-8927-7_3,

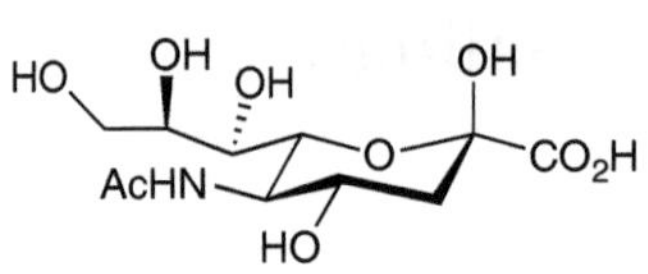

1

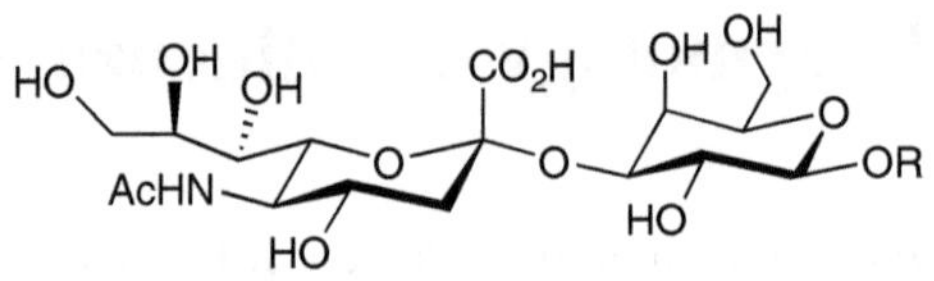

2

Fig. 1 Structure of β-sialic acid (**1**) and a generic structure for an α-sialosyl-(2→3)-β-galactoside (**2**)

[1]. Many sialidases possess more than one hydrolytic activity. For example, the influenza sialidase cleaves both 2,3- and 2,6-α-linked glycoconjugates, with the preferred regiochemistry for catalysis dictated by the strain of the virus. In general, the human influenza virus sialidase readily cleaves a Neu5Acα2,6Gal motif, the predominant linkage found on epithelial cells of the human trachea [2], while the corresponding enzyme from avian influenza virus preferentially recognizes and cleaves Neu5Acα2,3Gal linkages (**2**, Fig. 1) [3]. Substrate specificity exceptions also have been documented [4].

Despite their diverse origins, all *exo*-sialidases characterized to date have been shown to be retaining glycosidases [5, 6]. The initial product released by an *exo*-sialidase is α-sialic acid (*N*-acetylneuraminic acid, Neu5Ac), which upon mutarotation [7, 8] produces a mix of the thermodynamically favoured *β*-anomer (91.2%), the α-anomer (5.8%), and three acyclic species (the ketone, an enol, and a *gem*-diol) [9].

1.1 Biological Relevance of Sialidases

In addition to the *exo*-sialidases, a second class of glycosyl hydrolases, the *endo*-sialidases (EC 3.2.1.129), cleaves sialyl linkages located at internal positions of polysialic acid. For example, bacteriophage E expresses an *endo*-sialidase that cleaves 2,8-linked polysialic acid [10, 11]; recent work shows that enzyme catalysis occurs with inversion of configuration [12]. *exo*-Sialidases are recognized to play important roles in biologically relevant processes including cellular recognition and cancer metastasis [13, 14], and that they are linked to human diseases such as cholera and influenza [15, 16].

In the case of influenza, the virus particle contains two prominent surface glycoproteins, the influenza sialidase and haemagglutinin, a carbohydrate-binding protein that helps initiate the transport of the viral genome into the host cell [17]. Once a host cell is infected, the virus directs the production of new viral proteins and the replication of its genome. New viral particles assemble and bud off from the host cell, and the released viral progeny clump together and adhere to the host cell surface as a result of interactions between viral haemagglutinin and the sialylated host and viral glycoproteins [18]. The infectivity of virions contained in this adherent mass is compromised until the viral sialidase cleaves the surface sialic acid residues from the glycoproteins, thereby enabling the newly formed, functional virions to be released to begin a new round of infection. Thus, the critical role played by influenza sialidase in the virus life cycle makes this enzyme a primary target for the development of antiviral therapeutics that interfere with the release of infectious influenza virus progeny from infected host cells [19, 20].

1.2 Enzyme Kinetics and Catalysis

With derivation of the Michaelis–Menten equation for single substrate enzymes [21], kinetics has assumed a central role in elucidating the origins and causes of biological catalysis. Two key kinetic parameters relevant to a discussion of sialidase-catalysed reactions are k_{cat} (or turnover number), the first-order rate constant for the conversion of the Michaelis complex to free enzyme and product, and k_{cat}/K_m, the apparent second-order rate constant for the reaction of free enzyme and free substrate [22]. For enzymatic activity to be viewed in the proper context, one must appreciate the chemical challenges faced in catalysing the reaction of interest. For example, although the water-catalysed mutarotation of glucose is an intrinsically rapid reaction ($t_{½} = 28$ min in "pure" water) [23, 24]), nature produces the enzyme aldose 1-epimerase (EC 5.1.3.3; mutarotase) to accelerate this step in the metabolic processing of glucose, the major energy source for living beings. Of interest, an analogous enzymatic activity specific for sialic acid was recently discovered [25]. Both enzymatic efficiency (k_{cat}/K_m) and enzyme proficiency ($k_{cat}/K_m \times 1/k_{uncat}$) [26] must be considered to achieve a better understanding of biological catalysis (Fig. 2).

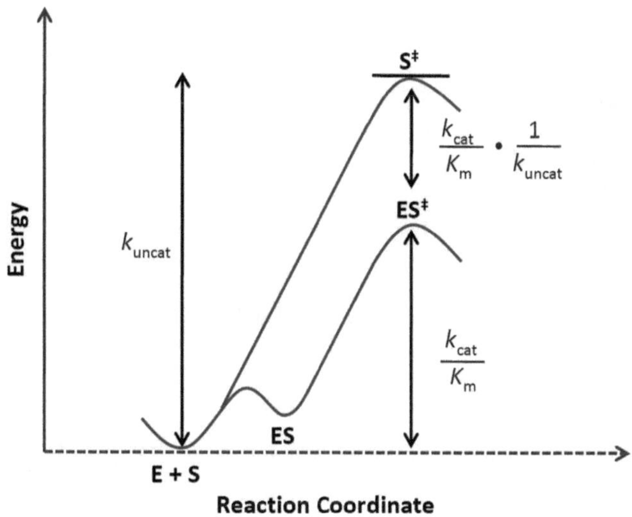

Fig. 2 Free energy diagram showing catalytic proficiency ($k_{cat}/K_m \times 1/k_{uncat}$) and catalytic efficiency (k_{cat}/K_m)

2 Mechanistic Tools

2.1 Linear Free Energy Relationships

In the context of biological catalysis, a useful Linear Free Energy Relationships (LFER) is a Brønsted plot, which graphs the logarithm of a rate constant against a pK_a value. For instance, a plot of $\log(k_{obs})$ against the pK_a value of the leaving group's conjugate acid can provide insightful mechanistic information about the degree of glycosidic bond cleavage at the reaction transition state (TS) [27, 28]. In solution, the spontaneous hydrolyses of aryl α- and β-sialosides (**3** and **4**, respectively; Fig. 3) give rise to β_{lg} values (Brønsted plot slopes) of −1.32 [29] and −1.22 [30], respectively. These large negative numbers can be considered to be the limiting values for reaction TS that do not have general acid or nucleophilic components. Similar β_{lg} values of −1.22 were measured for the spontaneous hydrolyses of the neutral and anionic forms of pyridinium α-sialosides (**5**, Fig. 4), thus lending support to glycosidic bond cleavage (C–N in this case) being the sole rate-determining step for the reaction [31]. The β_{lg} values for enzyme-catalysed reactions tend to be smaller than those for hydrolyses in solution, with three main factors accounting for the decrease: (1) general acid catalysis reduces the charge on the leaving group; (2) nucleophilic catalysis reduces the degree of glycosidic bond cleavage at the TS; and (3) a step other than glycosidic bond cleavage is rate determining.

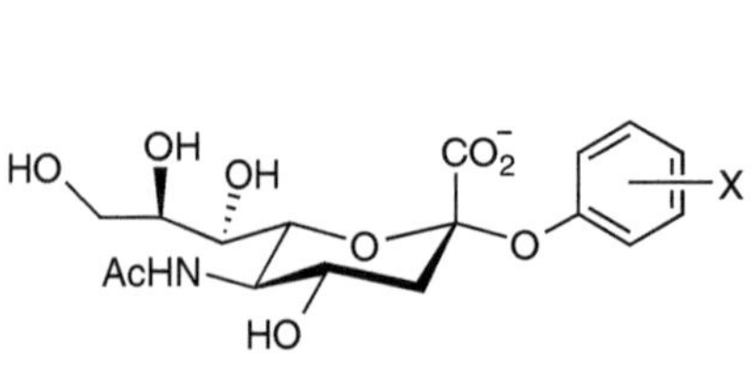

3

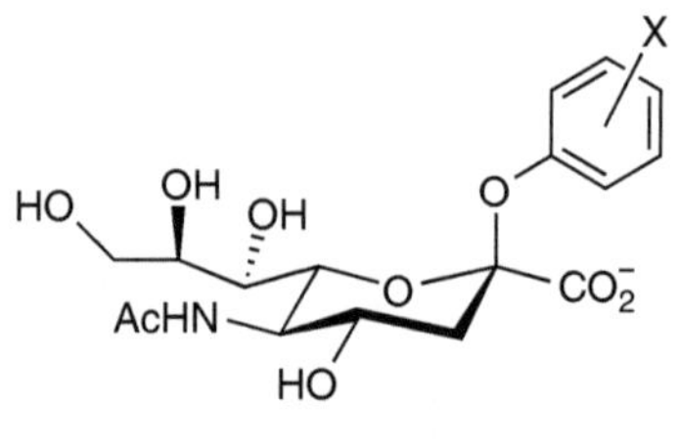

4

Fig. 3 Generic structures for aryl α-sialoside (**3**) and aryl β-sialoside (**4**)

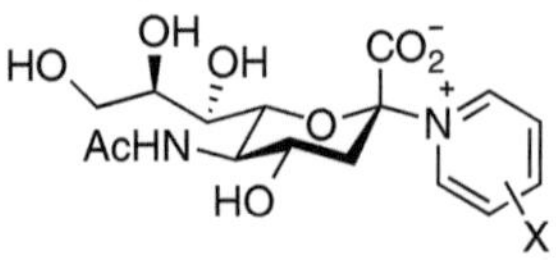

5

Fig. 4 Generic structure for α-sialosyl pyridinium salts (**5**)

2.2 Kinetic Isotope Effects

The measurement of Kinetic Isotope Effects (KIEs) is commonly used by chemists and biochemists as a mechanistic probe for enzymatic reactions. As an example, KIEs can be used to distinguish between concerted or step-wise mechanisms as well as to determine which bond forming and breaking processes are occurring at the relevant enzymatic TSs [32, 33].

The potential energy surface for chemical systems can be regarded as invariant for isotopomers and as such these species would have identical structures in a hypothetical vibrationless equilibrium state. For systems that contain many atoms, the major cause of an isotope effect is the zero-point energy (ZPE) difference that exists between light and heavy isotopes of the starting material and the reaction TS. As a rule of thumb, heavier isotopes will accumulate in positions of stronger bonding: if a bond is broken or weakened, the heavier isotopomer will have a larger free energy of activation, and thus, a lower rate of reaction (Fig. 5).

Isotopic substitution at positions that undergo bond cleavage or formation at the reaction TS gives rise to primary KIEs, and the magnitude of these effects is generally larger when the relative isotopic mass change is greatest. For heavy-atom primary KIEs such as those associated with carbon substitution (k_{12}/k_{13}), maximal values are small (generally less than 1.10) and, as such, are often difficult to determine accurately. Secondary KIEs are produced when the isotopic substitution is at a position not directly involved in formal bond changes at the TSs. Secondary deuterium KIEs (SDKIE) are classified as either α or β depending on whether the position of substitution is at or adjacent to the reaction centre, respectively. KIEs of this type are generally associated with bond re-hybridization and changes in hyperconjugation.

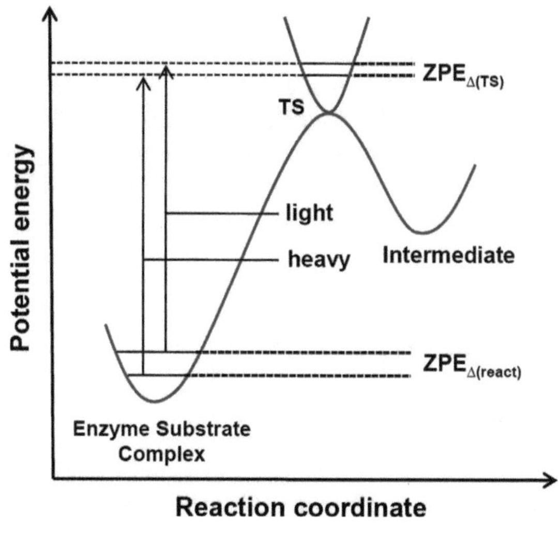

Fig. 5 Differential zero point energy differences in activation energy between heavy and light isotopomers

As the two kinetic parameters k_{cat} and k_{cat}/K_m give complementary information, it is often desirable to determine KIEs on both of these parameters. Typically, the most accurate KIE measurements are generated by competitive experiments in which labelled and unlabelled substrates are present simultaneously in the solution, an approach that minimizes systematic errors. Regardless of substrate concentration, these types of measurements only give KIE values on the k_{cat}/K_m parameter [32]. In contrast, determination of KIEs on k_{cat} only can be accomplished by a comparison of separately measured rate differences for labelled and unlabelled substrates [32].

2.3 *Comparisons Between Sialosides and Glucosides: Intrinsic Reactivity*

Activated sialyl-based ketal centres are approximately 10^4-fold more reactive than acetal-based glucopyranosides. For instance, using a β_{lg} value of -1.32 [29] for the uncatalysed hydrolysis of 4-nitrophenyl *N*-acetyl-α-D-neuraminide (PNP-Neu5Ac, **6** X = H, Fig. 6) and the reported rate constant for the reaction [29, 30], the estimated rate constant for spontaneous hydrolysis of 2,4-dinitrophenyl *N*-acetyl-α-D-neuraminide (**6** X = NO_2, Fig. 6) at 37°C is approximately 0.06 s^{-1}, whereas the corresponding hydrolysis of 2,4-dinitrophenyl β-D-glucopyranoside (**7**, Fig. 6) occurs with a rate constant of 5.6×10^{-6} s^{-1} [34]. This enhanced reactivity in aqueous solutions likely results from three main factors: (1) glucose possesses an electron-withdrawing hydroxyl group adjacent to the anomeric centre (C-2), while the corresponding position (C-3) in sialic acid is unsubstituted; (2) glucopyranosides contain an acetal-based anomeric centre, whereas sialosides incorporate a more sterically crowded ketal unit containing a carboxylate substituent that is in an axial (α-anomer) or an equatorial (β-anomer) orientation; and (3) the negatively charged carboxylate group can provide electrostatic stabilization of the adjacent positively charged oxacarbenium ion intermediates.

Of note, the non-enzymatic hydrolysis of *p*-nitrophenyl *N*-acetyl-α-D-neuraminide (PNP-Neu5Ac) provides an excellent example of the mechanistic complications introduced by a carboxylate group as compared to glucoside substrates. The solvent deuterium KIE obtained by Sinnott and co-workers for the

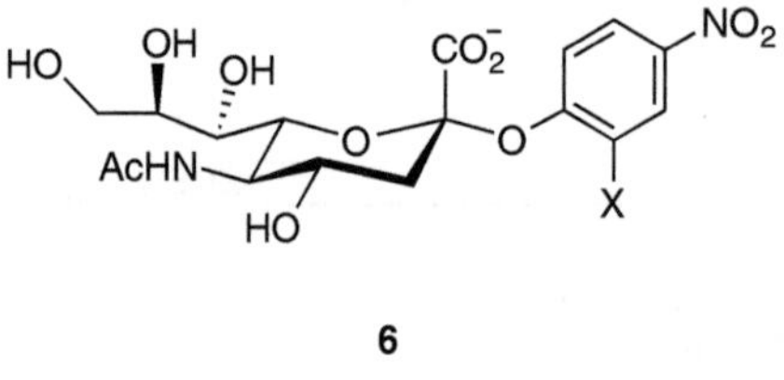

6

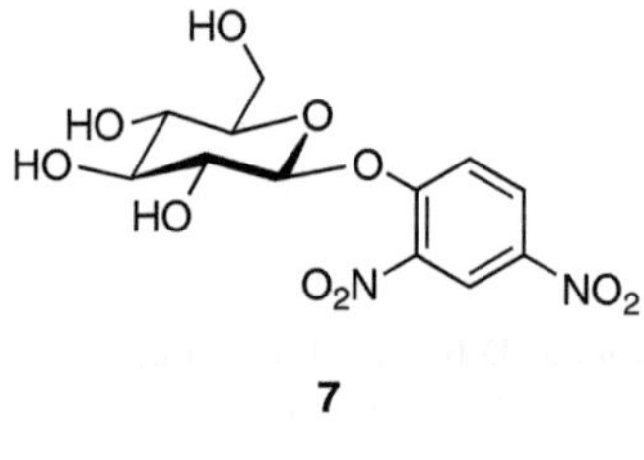

7

Fig. 6 Structures of PNP-Neu5Ac (**6**, X=H) and 2,4-dinitrophenyl β-D-glucopyranoside (**7**)

spontaneous reaction of PNP-Neu5Ac in solution led them to propose that cleavage of the glycosidic bond proceeds via nucleophilic attack by the C-1 carboxylate of the sialoside on the anomeric carbon atom (**8**, Fig. 7) [29].

However, all recent mechanistic studies on the non-enzymatic reactions in aqueous solutions of sialosides, including aryl α- and β-sialosides [30], pyridinium α-sialosides [31, 35], and CMP β-sialoside (**9**, Fig. 7) [36, 37], have concluded that the carboxylate group does not assist in the departure of the aglycone fragment.

2.4 *Oxacarbenium Ion Lifetimes*

Horenstein and co-workers studied the specific-acid catalysed and spontaneous hydrolysis reactions of CMP *N*-acetyl-β-D-neuraminic acid (**9**) [36, 37] and suggested that both pathways give rise to sialyl oxacarbenium ion intermediates. Based on results that included anomeric 13-C KIEs and product studies these researchers estimated that the sialyl oxacarbenium ion (**10**, Fig. 8) has a lifetime of greater than 3×10^{-11} s [37]. In contrast, the estimated lifetime of $1\text{–}3 \times 10^{-12}$ s for the quintessential carbohydrate cation the glucopyranosylium ion (**11**, Fig. 8) is at least ten times shorter [38, 39]. The absence of a hydroxyl group adjacent to the anomeric centre in the sialosyl system gives rise to a significant fraction of this effect, as oxacarbenium ions associated with acyclic analogues of 2-deoxy hexopyranose sugars have approximately fourfold longer lifetimes in water than the corresponding carbohydrate mimic [38].

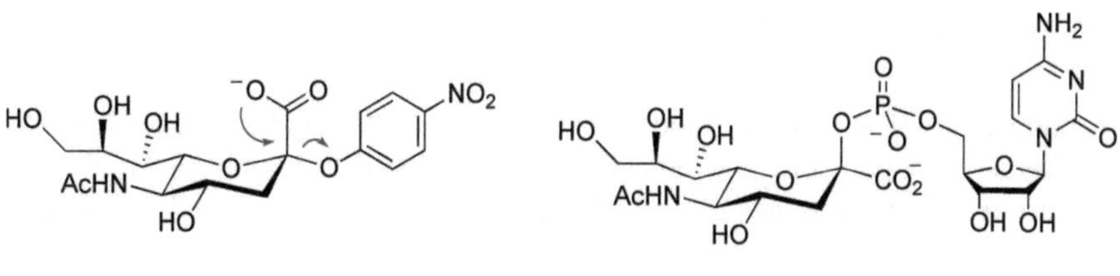

8 **9**

Fig. 7 Intramolecular nucleophilic attack in PNP-Neu5Ac (**8**) and the structure of CMP-sialic acid (**9**)

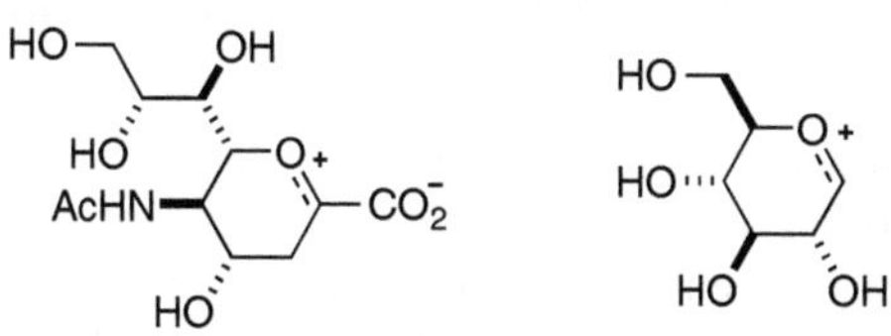

10 **11**

Fig. 8 Structure of the sialosyl oxacarbenium ion (**10**) and the glucopyranosylium ion (**11**)

3 Sialidase Catalysis

3.1 Sialidase Active Site Homology

Diverse origins notwithstanding, mammalian, bacterial, and viral sialidases share fundamental structural similarities. Three glycosyl hydrolase families contain *exo*-sialidases (GH 33 – bacterial and eukaryotic enzymes; GH 34 – influenza sialidases; and GH 83 – dual function haemagglutinin-neuraminidases enzymes) and all of these families belong to clan GH-E, which contains a sixfold β-propeller as the prominent structural motif. Moreover, the enzymatic active sites of each of these sialidases feature seven highly conserved amino acid residues [40]: a pair of acidic residues and a tyrosine residue are important for catalysis, while an arginine triad and a structural glutamate are also present [41]. The positively charged arginine triad mediates strong electrostatic interactions with the negatively charged carboxylate of the α-sialoside substrate. The importance of the triad was demonstrated by a mutagenesis study in which two active site arginines from the influenza sialidase, R152 and R371, were replaced with R152K/R152I and R371K, respectively [42]. Although correctly folded mutant enzymes were expressed, the sialidase activities were severely compromised. Both R152 mutants gave rise to inactive enzymes, and the R371 mutant displayed markedly reduced enzyme activity (5–10% of the wild-type activity) [42].

3.2 Catalytic Residues

In contrast to typical retaining glycosidases where two key acidic residues function as either a nucleophile or a general acid/base, crystal structure analysis of influenza sialidase reveals the absence of a suitably positioned carboxylate nucleophile [43, 44]. For sialidases, the key catalytic residues have been identified as a glutamate/tyrosine pair and an aspartic acid residue [45].

3.3 Stereochemical Outcome of Catalysed Reactions

An analogy can be drawn from the catalytic mechanism for retaining glycosidases to account for the observed stereochemical outcome of the sialidase-catalysed reaction. The generally accepted mode of action for most retaining glycosidases is a double displacement mechanism that produces an enzyme-bound intermediate [46, 47]. Both the formation (glycosylation) and subsequent breakdown (deglycosylation) of this critical intermediate are thought to involve ring-distorted oxacarbenium ion-like transition states (TSs; Fig. 9) [48, 49].

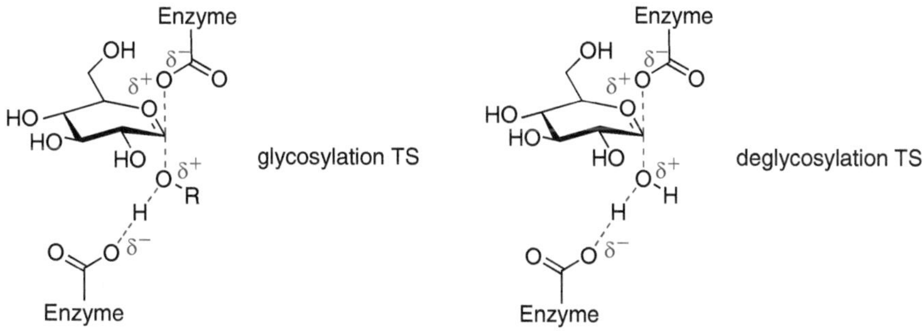

Fig. 9 Generic transition state structures for the glycosylation and deglycosylation reactions of retaining glycosidases

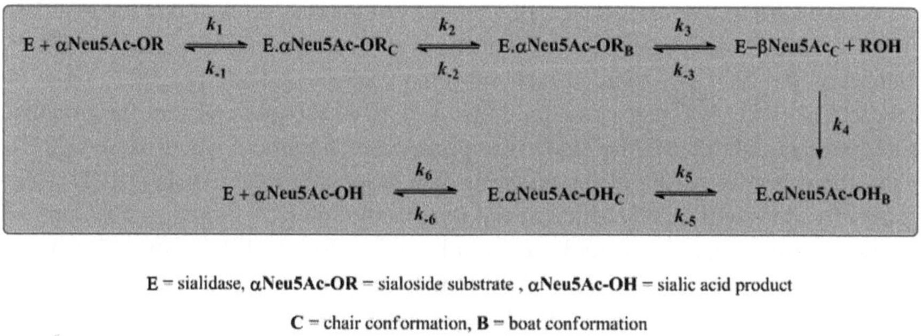

Fig. 10 Mechanistic scheme for sialidase-catalysed hydrolysis reactions

3.4 Catalytic Mechanism of Influenza Sialidase

The mechanistic scheme proposed in 1994 by Guo et al. for retaining sialidases has stood the test of time [50]. The key features of their detailed proposal (Fig. 10) involve formation of the Michaelis complex with the substrate bound to the enzyme in the normal $^2C_5$ chair conformation; conformational change of the enzyme: substrate complex to convert the substrate into a boat conformation; and subsequent cleavage of the glycosidic bond to give an enzyme-bound intermediate. Hydrolysis of this intermediate generates the sialic acid product bound to the enzyme in a boat conformation, and following another conformational change, the sialic acid dissociates from the active site.

Given that all *exo*-sialidases contain identical active site residues and that the *exo*-sialidase family belongs to a single glycosyl hydrolase clan, one can reasonably assume that these enzymes operate by the same overall mechanism and that family members differences in catalytic behaviour are likely restricted to the relative free energy levels of the various intermediates and TSs. Thus, throughout the discussion of the mechanism of action of influenza enzymes, examples and conclusions from other sialidases will be included, as appropriate.

3.5 Evidence for a Conformational Change of the Initial-Formed Michaelis Complex

In their landmark paper on the mechanism of action of an influenza sialidase, Sinnott and co-workers noted that the observed SDKIE on V/K: "... were not fully expressed at pH 6, in accord with the postulation of an isotope insensitive step preceding bond cleavage." In other words, a step that involves a conformational change of the first-formed Michaelis complex is partially rate determining at a pH value of 6. Similarly, Chou et al. argued that the rate-determining step for k_{cat}/K_m on the influenza sialidase-catalysed hydrolysis of non-natural pyridinium sialoside substrates (**5**) is a conjoint conformational change of the first-formed enzyme–substrate complex, and that this event is followed by rapid and irreversible glycosidic bond cleavage [31].

With regard to other sialidases, Bennet and co-workers showed that the sialidase from *M. viridifaciens* catalyses the hydrolysis of 4-nitrophenyl α-D-sialoside (PNP-Neu5Ac, **6**, X = H) with a second-order rate constant ($k_{cat}/K_m = 1.7 \times 10^7\ M^{-1}\ s^{-1}$; [51, 52]) that is within a factor of 10 of the rate constant for enzymes such as acetylcholinesterase and triosephosphate isomerase which have k_{cat}/K_m values that border on the diffusion-controlled association rate constant [22]. Consequently, this catalytically efficient sialidase cannot directly bind the substrate in a boat conformation, because to do so would require close to 10% of the substrate to be in this high-energy form. In addition, based on the observed reactivity of sialosides and thiosialosides with wild-type and active site mutant *M. viridifaciens* sialidases, Narine et al. concluded that an obligatory conformational change of the Michaelis complex is a kinetically important feature of the mechanism of action for this bacterial sialidase [51].

3.6 Identity of the Enzyme-Bound Intermediate – Covalent or Cationic?

Figure 11 shows the three possibilities that have been considered for the identity of the sialidase-bound intermediate, namely: (1) an α-lactone formed transiently by an intramolecular nucleophilic attack of the substrate's carboxylate group; (2) an

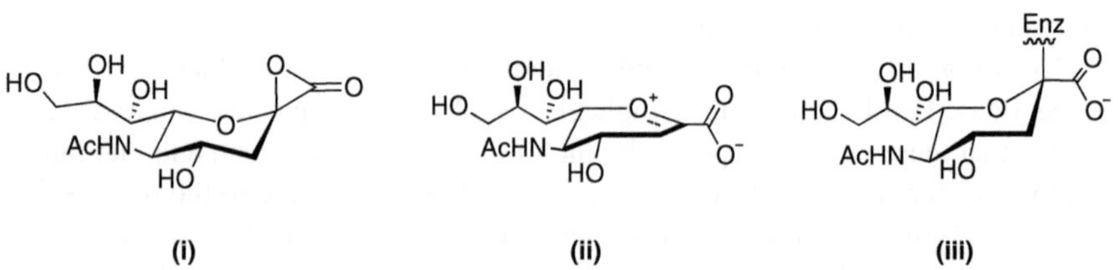

(i) (ii) (iii)

Fig. 11 Structures for the three possible sialoside-enzyme bound intermediates

oxacarbenium ion, which is stabilized by the adjacent glutamate/tyrosine pair; and (3) a covalent glycosyl enzyme species, which could have been formed via either an S_N1 or an S_N2-type reaction.

3.7 *Intramolecular Nucleophilic Attack*

Based on their original analysis of the non-enzymatic reactions of sialosides, Sinnott and co-workers suggested that the C-1 carboxylate of the sugar might have a catalytic function during sialidase-catalysed hydrolysis reactions [50]. This idea was founded in part on the belief that the enzyme active site glutamate is unsuitably positioned to function as a nucleophile [50]. The substrate's carboxylate group is, in fact, anchored in position by strong electrostatic interactions to three positively charged arginine residues, as shown by the average distance of 3.18 Å for the six closest O–N contacts for sialic acid bound to influenza sialidase [44]. Therefore, it is unlikely that this group can act as an intramolecular nucleophile, and indeed, Sinnott and co-workers favoured a conformational change of the initial-formed enzyme:product complex as the rate-determining step for k_{cat} [50].

3.8 *Enzyme-Bound Oxacarbenium Ion Intermediate*

A small but reproducible observed inverse β-SDKIE on k_{cat}/K_m of 0.979 led Chong et al. to propose that influenza sialidase-catalysed hydrolysis gives a sialosyl cation intermediate [53]. Both this work and an independent KIE study by Guo et al. [50], which reported a KIE of close to 1.0, employed separate reaction rate measurements for the various isotopomers, an approach that can introduce small systematic errors in the data. In contrast, the experimental procedure for measuring competitive KIEs on k_{cat}/K_m does not have this inherent problem.

Several theoretical studies on sialidase catalysis have suggested that these reactions proceed via a cationic oxacabenium ion intermediate bound to the active site [54, 55]. Many of these reports implicitly assume that the glycal inhibitor 5-acetamido-2, 6-anhydro-3,5-dideoxy-D-*glycero*-D-*galacto*-non-2-enoic acid (DANA, **12**, Fig. 12)

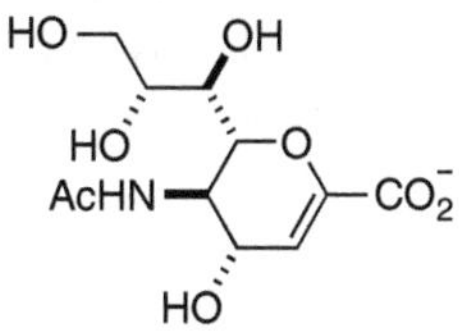

12

Fig. 12 Structure for 5-acetamido-2,6-anhydro-3,5-dideoxy-D-*glycero*-D-*galacto*-non-2-enoic acid (DANA, **12**)

is a good mimic of the reaction TS [55]. To date, however, there are no sialidase experiments to support a correlation between inhibitor potency and either catalytic efficiency or proficiency, a requirement for a true TS analogue inhibitor [56].

3.9 Covalent Enzyme-Bound Intermediate

The fluorine-substituted glycoside methodology pioneered by Withers and co-workers is an elegant chemical tool used to probe for potential active site nucleophilic residues [57]. The method generates rate reductions via the destabilization of oxacarbenium ion-like TSs formed during both glycosylation and deglycosylation steps by employing substrates that contain an electronegative fluorine atom. The retardation of positive charge development during glycosylation can be ameliorated by a highly activated substrate containing a good leaving group such as fluoride or 2,4-dinitrophenolate; as a result, deglycosylation of the enzyme-bound intermediate often becomes the rate determining step of the reaction. In the current context, Watts et al. showed that the 2,3-difluoro sialic acid analogue, **13** (Fig. 13) efficiently trapped the *Trypanosoma cruzi trans*-sialidase (TcTS) [58]. *T. cruzi* is an example of a unicellular pathogenic protozoan that is unable to biosynthesize sialic acid [59]. The parasite produces *trans*-sialidase enzymes (GH family 33) that are effective at scavenging sialic acid from external sources. Although *trans*-sialidases are members of the sialidase superfamily, the enzyme preferentially catalyses the transfer, with retention of configuration, of terminal α-(2,3) linked sialic acid from galactose residues on host glycoconjugates to terminal galactose residues of glycoconjugate chains on the external surface of the parasite [60]. As expected, these *trans*-sialidases also feature the same seven highly conserved active site residues that characterize members of the *exo*-sialidase superfamily.

Watts et al. showed by ES/MS analysis that incubation of **13** with TcTS resulted in a mass increase of 304 ± 12 Da, an observation that points to the formation of a covalently bound sialyl-enzyme intermediate. Indeed, following peptide digestion of both labelled and unlabelled TcTS proteins, LC–MS analysis of the fragment peptides confirmed that the active site tyrosine had been covalently modified [58].

Two ensuing experiments demonstrated the mechanistic relevance of the trapped sialyl-enzyme intermediate [58]. In the first of these key studies, labelled enzyme incubated with the competent acceptor lactose **14** (Fig. 13) exhibited transferase activity, while the second showed there to be no turnover of the intermediate when

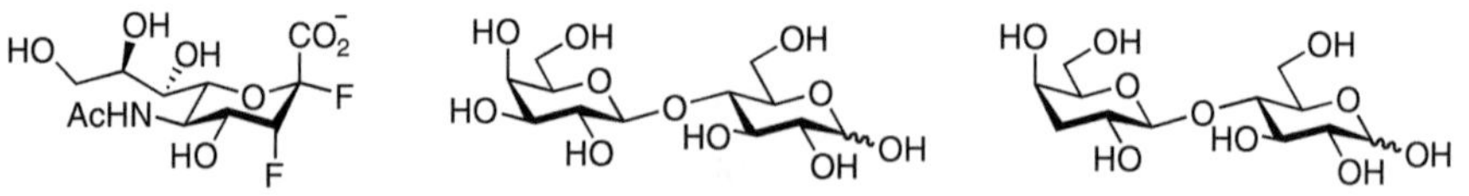

Fig. 13 Structures of 2,3-difluorosialic acid analogue (**13**), lactose (**14**), and 3'-deoxy-lactose (**15**)

the acceptor was changed to 3'-deoxy lactose **15** (Fig. 13). Thus, Watts et al. concluded that TcTS operates via a double displacement mechanism involving transient formation of a covalent intermediate, with the active site conserved tyrosine residue acting as the nucleophile.

Further work used the difluorosialic acid trapping methodology coupled with single crystal X-ray diffraction studies to show that the sialyl-tyrosinyl intermediate formed by the reaction of **13** with various sialidases and *trans*-sialidases was in a $^{2}C_{5}$ chair conformation [61–63].

In a separate mutagenesis study, Bennet and co-workers changed the conserved active site tyrosine Y370 of the *Micromonospora viridifaciens* sialidase to an alanine (Y370A), a glycine (Y370G) and an aspartic acid (Y370D) [52]. As determined by NMR spectroscopic analysis, the reaction mechanism for each of the resulting mutant enzymes had changed from retention to inversion of configuration at the anomeric centre. An interpretation of this result is that when tyrosine is substituted with smaller amino acid residues, a cavity forms in the enzyme active site, and it can accommodate one or more water molecules even in the presence of bound substrate. In this environment, a water molecule is appropriately positioned to act as a nucleophile for either the direct displacement of the leaving group or the capture of a newly formed oxacarbenium ion intermediate, processes which both generate an inverted hydrolysis product [64].

3.10 Glycosylation Transition State Structure

Given that most experimental evidence points to the involvement of the covalently bound enzyme intermediate being a catalytically relevant species, a pertinent question is whether this intermediate is formed via an associative (S_N2 – A_ND_N, concurrent nucleophilic attack) or a dissociative (S_N1 – $D_N * A_N$, to give a short lived oxacarbenium ion) TS. Clearly, in aqueous solution the reaction mechanism of glycopyranosides is finely balanced between these two alternatives [65]. Substitution of the C-2 hydroxyl group for a fluorine atom in both mannose and glucose likely changes the mechanism in solution from S_N1 to S_N2 by reducing the lifetime of the oxacarbenium ion below that of a viable intermediate; as a result, the 2-deoxy-2-fluoro-glycopyranosides react via an enforced concerted or a pre-association mechanism [66]. This analysis is consistent with the observed positive rate constant deviations of 2-deoxy-2-fluoroglycosides from the correlations for 2-deoxy and parent carbohydrates that are noted in the mechanism paper of Withers and co-workers [34]. Undoubtedly, the lifetime of the 3-fluorosialosyl oxacarbenium ion (**16**, Fig. 14) will be shorter than that of the parent cation (**10**, Fig. 8); however, it has yet to be determined if this substitution results in significant perturbations to the mechanism of reaction either in solution or for the sialidase-catalysed event.

The best evidence to date that sialidases catalyse reactions via associative TS (S_N2) can be found in a heavy atom KIE study on the *T. cruzi trans*-sialidase by

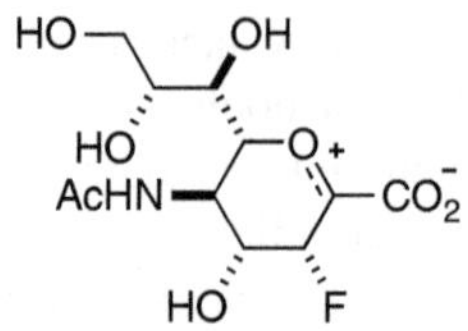

16

Fig. 14 Structure of the 3-fluoro-sialosyl oxacarbenium ion (**16**)

Horenstein and co-workers [67]. In this report, the anomeric carbon KIE (k_{12}/k_{13}) on k_{cat}/K_m is 1.032 for the transfer of a sialic acid residue from the slow substrate sialyl-galactose onto a lactose acceptor. The magnitude of this effect is clearly within the expected range (k_{12}/k_{13} ~ 1.03–1.08) for an associative mechanism (S_N2) but outside of the normal limits for a dissociative substitution reaction (S_N1, k_{12}/k_{13} ~ 1.00–1.01) [32]. Given that cleavage of the anomeric sialic acid-galactose bond is rate determining, the magnitude of the reaction centre KIE is consistent with the occurrence of a simultaneous nucleophilic attack, which based on the trapping and mutagenesis experiments noted earlier involves the active site tyrosine residue, i.e. an A_ND_N (S_N2) reaction.

3.11 Transition State Geometry

The large SDKIE at pH 6.0 of 1.09 ± 0.02 for the dideuterated 4-methylumbelliferyl (3-^{2} H)$_2$-*N*-acetyl-α-D-neuraminic acid on k_{cat}/K_m suggests that the influenza sialidase-catalysed TS involves a significant degree of hyperconjugation between the C–D bonds and the nascent positive charge at the anomeric centre [53]. Moreover, KIEs measured using singly labelled *R*-(3-^{2} H) and *S*-(3-^{2} H) PNP-Neu5Ac are consistent with a TS conformation that is similar to a $B_{2,5}$ boat [50], likely a ^{4} H_5 half chair [47].

3.12 Rationale for Tyrosine as an Active Site Nucleophile

It has been suggested that potentially unfavourable electrostatic interactions between the substrate carboxylate and the customary nucleophile – a negatively charged aspartate or glutamate group – have directed the evolution of the neutral nucleophile tyrosine [58]. This explanation is unlikely however, considering that, as noted earlier, the carboxylate of the sialoside substrate is closely surrounded by three positively charged arginine residues. A second possibility is that the intrinsically greater reactivity of sialosides requires a sialosyl-enzyme intermediate to have a worse leaving group (a tyrosine, rather that a carboxylate) than a normal glycosidase to allow enough time for the carbohydrate leaving group to diffuse from the active site and be replaced by water prior to breakdown of the intermediate [52].

3.13 Intrinsic Reactivity and Catalysis

In summary, sialidases possess three complementary strategies that combine to provide efficient hydrolysis of sialosides. Coming into play first is the strong binding of the enzyme via electrostatic interactions (*vide supra*) with the substrate's carboxylate group and an interaction between a hydrophobic pocket on the enzyme that binds the substrate's amide methyl group [68]. Of added importance, these two interactions occur at opposite sides of the substrate's pyranosyl ring and likely function as "levers" to drive enzyme-promoted conformational changes to the carbohydrate during the catalytic cycle. As a second strategy, the enzyme uses nucleophilic catalysis by the active site tyrosine residue and this is likely to proceed with general base assistance from the neighbouring glutamate residue. Lastly, the enzyme employs electrophilic catalysis from the active site aspartic acid residue.

Since the advent of site-directed mutagenesis, enzymologists have used this technique to identify amino acid residues that are important in enzyme catalysis. Experiments using sialidases altered at the active site tyrosine clearly indicate that that amino acid residue is critical for nucleophilic catalysis [69–71]. When the activity of mutant sialidases is compared to that of wild-type enzymes, Brønsted plot analyses give β_{lg} values that are more negative for enzymes that have had their active site tyrosine replaced by a poorly nucleophilic residue such as phenylalanine [72] or a small amino acid residue such as glycine or aspartic acid, which provides the environment for a bound water molecule to act as the nucleophile [52]. In general, these sialidase mutants show much smaller decreases in catalytic activity relative to the parent enzyme when they are provided with activated substrates, such as PNP-Neu5Ac, and more dramatic decreases in reactivity with natural substrates

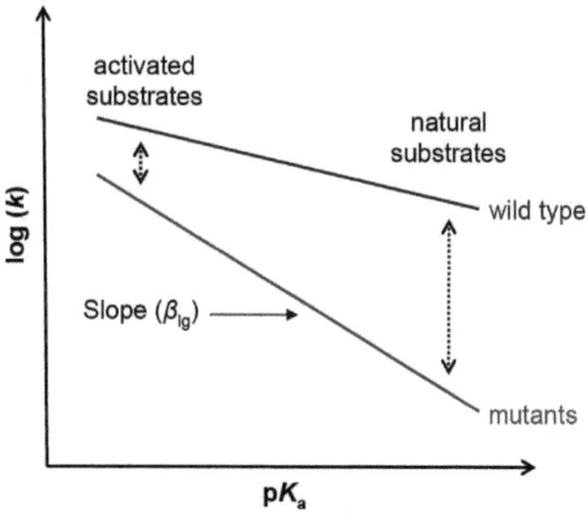

Fig. 15 Brønsted plot showing the larger negative β_{lg} values generally associated with mutant sialidases

[52, 72, 73]. The Brønsted plot shown in Fig. 15 compares wild-type and mutant catalysed reactions, and it illustrates, via a more negative β_{lg} value for the mutant-catalysed reactions, the greater reaction rate difference that occurs for reactions conducted with activated versus natural substrates.

The decrease in activity observed for sialidase active site tyrosine mutants is much smaller than the reduction seen for mutagenized retaining glycosidases lacking an active site carboxylate nucleophile. For example, the *Agrobacterium* β-galactosidase nucleophilic mutant (Glu358Ala) lost $>10^7$ of its catalytic power when the highly activated 2,4-dinitrophenyl β-D-galactoside was used as substrate [74], while the Y370F mutant *M. viridifaciens* sialidase is only about 100-fold less active with activated substrates [72].

The same general phenomenon is observed for sialidase mutants of the conserved aspartic acid residue [75], i.e. these mutants exhibit small reductions in the reaction rate when activated substrates are used and larger decreases with natural substrates (Fig. 15). In other words, the aspartic acid residue contributes weak general acid catalysis when the substrate contains an activated leaving group. Indeed, based on a measured leaving group ^{18}O-KIE for the *Vibrio cholorae* sialidase-catalysed hydrolysis of PNP-Neu5Ac, Sinnott and co-workers reached a similar conclusion [76].

The third catalytic residue for sialidases is the active site glutamate, which presumably acts as a general base to assist the adjacent tyrosine nucleophile during formation of the enzyme-bound intermediate. Little is known about the catalytic consequences of removing this residue; however, based on the observations noted earlier for both tyrosine and aspartic acid mutant sialidases, one would expect to see a larger rate reduction for natural substrates.

4 Conclusions

Kinetic and structural data has provided the basis for the following currently accepted mechanism for influenza sialidase-catalysed hydrolysis reactions (Fig. 16): (1) an α-sialoside substrate binds to the enzyme active site in a ground

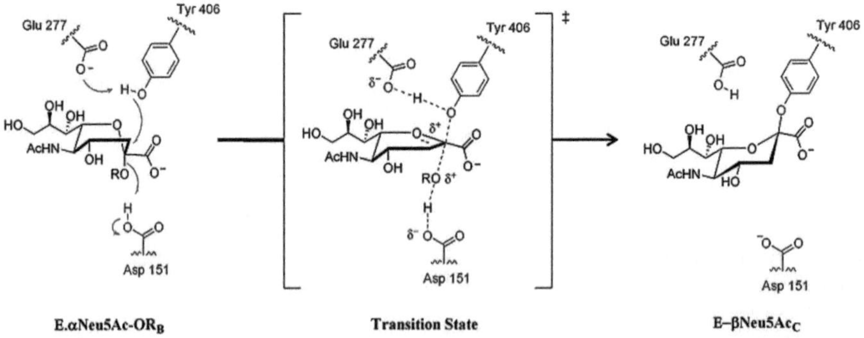

Fig. 16 Mechanism for the sialidase-catalysed formation of the covalently linked tyrosinyl β-sialoside enzyme bound intermediate

state $^2C_5$ chair conformation; (2) the enzyme–substrate complex undergoes a conformational change to give a species that is in a $B_{2,5}$-like boat conformation; (3) the active site tyrosine functions as a nucleophile to attack the anomeric centre of the sugar to generate a covalently bound sialosyl-enzyme intermediate in a $^2C_5$ chair conformation; (4) hydrolysis of this sialosyl-enzyme intermediate yields an α-sialic acid product bound to the enzyme in a $B_{2,5}$ boat conformation; and (5) a second conformational change is followed by diffusion of the product out of the active site to complete the catalytic cycle.

References

1. Saito M, Yu RK (1995) Biochemistry and function of sialidases. In: Rosenberg A (ed) Biology of the sialic acids. Plenum Press, New York, pp 261–313
2. Nelson J, Couceiro SS, Paulson JC, Baum LG (1993) Influenza-virus strains selectively recognize sialyloligosaccharides on human respiratory epithelium – the role of the host-cell in selection of hemagglutinin receptor specificity. Virus Res 29:155–165
3. Wilson JC, von Itzstein M (2003) Recent strategies in the search for new anti-influenza therapies. Curr Drug Targets 4:389–408
4. de Barros JF, Alviano DS, da Silva MH, Wigg MD, Alviano CS, Schauer R, Couceiro J (2003) Characterization of sialidase from an influenza A (H3N2) virus strain: kinetic parameters and substrate specificity. Intervirology 46:199–206
5. Wilson JC, Angus DI, von Itzstein M (1995) ^{1}H NMR evidence that *Salmonella typhimurium* sialidase hydrolyzes sialosides with overall retention of configuration. J Am Chem Soc 117: 4214–4217
6. Davies G, Sinnott ML, Withers SG (1998) Glycosyl Transfer. In: Sinnott ML (ed) Comprehensive biological catalysis. Academic, San Diego, CA, pp 119–209
7. Friebolin H, Supp M, Brossmer R, Keilich G, Ziegler D (1980) ^{1} H-NMR investigations on the mutarotation of *N*-acetyl-D-neuraminic acid. Angew Chem Int Ed Engl 19:208–209
8. Friebolin H, Kunzelmann P, Supp M, Brossmer R, Keilich G, Ziegler D (1981) ^{1} H-NMR-spekroskopische untersuchungen zur mutarotation der *N*-acetyl-D-neuraminsäure - pH-abhängigkeit der mutarotationsgeschwindigkeit. Tetrahedron Lett 22:1383–1386
9. Klepach T, Carmichael I, Serianni AS (2008) ^{13}C-Labeled *N*-acetyl-neuraminic acid in aqueous solution: Detection and quantification of acyclic keto, keto hydrate, and enol forms by ^{13}C NMR spectroscopy. J Am Chem Soc 130:11892–11900
10. Stummeyer K, Dickmanns A, Muhlenhoff M, Gerardy-Schahn R, Ficner R (2005) Crystal structure of the polysialic acid-degrading endosialidase of bacteriophage K1F. Nat Struct Mol Biol 12:90–96
11. Pelkonen S, Pelkonen J, Finne J (1989) Common cleavage pattern of polysialic acid by bacteriophage endosialidases of different properties and origins. J Virol 63:4409–4416
12. Morley TJ, Willis LM, Whitfield C, Wakarchuk WW, Withers SG (2009) A new sialidase mechanism: Bacteriophage K1F endosialidase is an inverting glycosidase. J Biol Chem 284: 17404–17410
13. Schauer R (1985) Sialic acids and their role as biological masks. Trend Biochem Sci 10: 357–360
14. Miyagi T, Kato K, Ueno S, Wada T (2004) Aberrant expression of sialidase in cancer. Trends Glycosci Glycotechnol 16:371–381
15. Corfield T (1992) Bacterial sialidases: roles in pathogenicity and nutrition. Glycobiology 2: 509–521
16. Taylor G (1996) Sialidases: Structures, biological significance and therapeutic potential. Curr Opin Struct Biol 6:830–837

17. Klenk H-D, Rott R (1988) The molecular biology of influenza virus pathogenicity. In: Maramorosch K, Murphy FA, Shatkin AJ (eds) Advances in virus research. Academic, New York, pp 247–281
18. Palese P, Tobita K, Ueda M, Compans RW (1974) Characterization of temperature sensitive influenza virus mutants defective in neuraminidase. Virology 61:397–410
19. von Itzstein M (2007) The war against influenza: discovery and development of sialidase inhibitors. Nat Rev Drug Discov 6:967–974
20. Islam T, von Itzstein M (2007) Anti-influenza drug discovery: Are we ready for the next pandemic? Adv Carbohydr Chem Biochem 61:293–352
21. Michaelis L, Menten ML (1913) Die kinetik der invertinwirkung. Biochem Zeit 49:333–369
22. Fersht A (1985) Enzyme structure and mechanism, 2nd edn. W.H. Freeman, New York
23. Hudson CS (1907) The catalysis by acids and bases of the mutarotation of glucose. J Am Chem Soc 29:1571–1576
24. Lewis BE, Choytun N, Schramm VL, Bennet AJ (2006) Transition states for glucopyranose interconversion. J Am Chem Soc 128:5049–5058
25. Severi E, Muller A, Potts JR, Leech A, Williamson D, Wilson KS, Thomas GH (2008) Sialic acid mutarotation is catalyzed by the Escherichia coli beta-propeller protein YjhT. J Biol Chem 283:4841–4849
26. Albery WJ, Knowles JR (1976) Evolution of enzyme function and development of catalytic efficiency. Biochemistry 15:5631–5640
27. Sinnott M (1998) Comprehensive biological catalysis: A mechanistic reference. Academic, San Diego,CA
28. Sinnott ML (2007) Carbohydrate chemistry and biochemistry: structure and mechanism. RSC Publishing, Cambridge
29. Ashwell M, Guo X, Sinnott ML (1992) Pathways for the hydrolysis of glycosides of *N*-acetylneuraminic acid. J Am Chem Soc 114:10158–10166
30. Dookhun V, Bennet AJ (2005) Unexpected stability of aryl β-*N*-acetylneuraminides in neutral solution: Biological implications for sialyl transfer reactions. J Am Chem Soc 127:7458–7465
31. Chou DTH, Watson JN, Scholte AA, Borgford TJ, Bennet AJ (2000) Effect of neutral pyridine leaving groups on the mechanisms of influenza type A viral sialidase-catalyzed and the spontaneous hydrolysis reactions of α-D-*N*-acetylneuraminides. J Am Chem Soc 122: 8357–8364
32. Melander LCS, Saunders WHJ (1980) Reaction rates of isotopic molecules. Wiley, New York
33. Kohen A, Limbach H-H (2006) Isotope effects in chemistry and biology. Taylor & Francis, Boca Raton, FL
34. Namchuk MN, McCarter JD, Becalski A, Andrews T, Withers SG (2000) The role of sugar substituents in glycoside hydrolysis. J Am Chem Soc 122:1270–1277
35. Knoll TL, Bennet AJ (2004) Aqueous methanolysis of an α-D-*N*-acetylneuraminyl pyridinium zwitterion: Solvolysis occurs with no intramolecular participation of the anomeric carboxylate group. J Phys Org Chem 17:478–482
36. Horenstein BA, Bruner M (1996) Acid-catalyzed solvolysis of CMP-N-acetyl neuraminate: evidence for a sialyl cation with a finite lifetime. J Am Chem Soc 118:10371–10379
37. Horenstein BA, Bruner M (1998) The *N*-acetyl neuraminyl oxecarbenium ion is an intermediate in the presence of anionic nucleophiles. J Am Chem Soc 120:1357–1362
38. Amyes TL, Jencks WP (1989) Lifetimes of oxocarbenium ions in aqueous solution from common ion inhibition of the solvolysis of α-azido ethers by added azide ion. J Am Chem Soc 111:7888–7900
39. Huang X, Surry C, Hiebert T, Bennet AJ (1995) The hydrolysis of 2-deoxy-β-D-glucopyranosyl pyridinium salts. J Am Chem Soc 117:10614–10621
40. Vimr ER (1994) Microbial sialidases: does bigger always mean better? Trends Microbiol 2:271–277
41. Henrissat B, Bairoch A (1996) Updating the sequence-based classification of glycosyl hydrolases. Biochem J 316:695–696

42. Lentz MR, Webster RG, Air GM (1987) Site-directed mutation of the active site of influenza neuraminidase and implications for the catalytic mechanism. Biochemistry 26:5351–5358
43. Varghese JN, Colman PM (1991) Three-dimensional structure of the neuraminidase of influenza virus A/Tokyo/3/67 at 2.2 A resolution. J Mol Biol 221:473–486
44. Varghese JN, McKimm-Breschkin JL, Caldwell JB, Kortt AA, Colman PM (1992) The structure of the complex between influenza virus neuraminidase and sialic acid, the viral receptor. Proteins: Struct Funct Genet 14:327–332
45. Colman PM, Varghese JN, Laver WG (1983) Structure of the catalytic and antigenic sites in influenza virus neuraminidase. Nature 303:41–44
46. Koshland DE Jr (1953) Stereochemistry and the mechanism of enzymic reactions. Biol Rev Camb Philos Soc 28:416–436
47. Vocadlo DJ, Davies GJ (2008) Mechanistic insights into glycosidase chemistry. Curr Opin Chem Biol 12:539–555
48. Davies GJ, Ducros VMA, Varrot A, Zechel DL (2003) Mapping the conformational itinerary of β-glycosidases by X-ray crystallography. Biochem Soc Trans 31:523–527
49. Davies GJ, Gloster TM, Henrissat B (2005) Recent structural insights into the expanding world of carbohydrate-active enzymes. Curr Opin Struct Biol 15:637–645
50. Guo X, Laver WG, Vimr E, Sinnott ML (1994) Catalysis by two sialidases with the same protein fold but different stereochemical courses: a mechanistic comparison of the enzymes from influenza A virus and *Salmonella typhimurium*. J Am Chem Soc 116:5572–5578
51. Narine AA, Watson JN, Bennet AJ (2006) Mechanistic requirements for efficient enzyme-catalyzed hydrolysis of thiosialosides. Biochemistry 45:9319–9326
52. Watson JN, Dookhun V, Borgford TJ, Bennet AJ (2003) Mutagenesis of the conserved active-site tyrosine changes a retaining sialidase into an inverting sialidase. Biochemistry 42:12682–12690
53. Chong AKJ, Pegg MS, Taylor NR, von Itzstein M (1992) Evidence for a sialosyl cation transition-state complex in the reactivity of sialidase from influenza virus. Eur J Biochem 207:335–343
54. Barnes JA, Williams IH (1996) Quantum mechanical/molecular mechanical approaches to transition state structure: mechanism of sialidase action. Biochem Soc Trans 24:263–268
55. Thomas A, Jourand D, Bret C, Amara P, Field MJ (1999) Is there a covalent intermediate in the viral neuraminidase reaction? A hybrid potential free-energy study. J Am Chem Soc 121: 9693–9702
56. Mader MM, Bartlett PA (1997) Binding energy and catalysis: the implications for transition-state analogs and catalytic antibodies. Chem Rev 97:1281–1301
57. Withers SG, Rupitz K, Street IP (1988) 2-Deoxy-2-fluoro-D-glycosyl fluorides. A new class of specific mechanism-based glycosidase inhibitors. J Biol Chem 263:7929–7932
58. Watts AG, Damager I, Amaya ML, Buschiazzo A, Alzari P, Frasch AC, Withers SG (2003) *Trypanosoma cruzi trans*-sialidase operates through a covalent sialyl-enzyme intermediate: tyrosine is the catalytic nucleophile. J Am Chem Soc 125:7532–7533
59. Colli W (1993) Trans-sialidase – a unique enzyme-activity discovered in the protozoan *Trypanosoma-cruzi*. FASEB J 7:1257–1264
60. Parodi AJ, Pollevick GD, Mautner M, Buschiazzo A, Sanchez DO, Frasch ACC (1992) Identification of the gene(s) coding for the *trans*-sialidase of *Trypanosoma cruzi*. EMBO J 11:1705–1710
61. Amaya MF, Watts AG, Damager T, Wehenkel A, Nguyen T, Buschiazzo A, Paris G, Frasch AC, Withers SG, Alzari PM (2004) Structural insights into the catalytic mechanism of *Trypanosoma cruzi trans*-sialidase. Structure 12:775–784
62. Watts AG, Oppezzo P, Withers SG, Alzari PM, Buschiazzo A (2006) Structural and kinetic analysis of two covalent sialosyl-enzyme intermediates on *Trypanosoma rangeli* sialidase. J Biol Chem 281:4149–4155
63. Newstead SL, Potter JA, Wilson JC, Xu GG, Chien CH, Watts AG, Withers SG, Taylor GL (2008) The structure of *Clostridium perfringens* NanI sialidase and its catalytic intermediates. J Biol Chem 283:9080–9088

64. Newstead S, Watson JN, Knoll TL, Bennet AJ, Taylor G (2005) Structure and mechanism of action of an inverting mutant sialidase. Biochemistry 44:9117–9122
65. Bennet AJ, Kitos TE (2002) Mechanisms of glycopyranosyl and 5-thioglycopyranosyl transfer reactions in solution. J Chem Soc Perkin Trans 2:1207–1222
66. Jencks WP (1980) When Is an intermediate not an intermediate? Enforced mechanisms of general acid-base catalyzed, carbocation, carbanion and ligand exchange reactions. Acc Chem Res 13:161–169
67. Yang J, Schenkman S, Horenstein BA (2000) Primary ^{13}C and beta-secondary ^{2}H KIEs for *trans*-sialidase. A snapshot of nucleophilic participation during catalysis. Biochemistry 39:5902–5910
68. Crennell SJ, Garman EF, Philippon C, Vasella A, Laver WG, Vimr ER, Taylor GL (1996) The structures of *Salmonella typhimurium* LT2 neuraminidase and its complexes with three inhibitors at high resolution. J Mol Biol 259:264–280
69. Ghate AA, Air GM (1998) Site-directed mutagenesis of catalytic residues of influenza virus neuraminidase as an aid to drug design. Eur J Biochem 58:320–331
70. Kleineidam RG, Kruse S, Roggentin P, Schauer R (2001) Elucidation of the role of functional amino acid residues of the small sialidase from *clostridium perfringens* by site-directed mutagenesis. Biol Chem 382:313–319
71. Wang Y, Yamaguchi K, Shimada Y, Zhao XJ, Miyagi T (2001) Site-directed mutagenesis of human membrane-associated ganglioside sialidase – Identification of amino-acid residues contributing to substrate specificity. Eur J Biochem 268:2201–2208
72. Watson JN, Newstead S, Narine A, Taylor G, Bennet AJ (2005) Two nucleophilic mutants of the *Micromonospora viridifaciens* sialidase operate with retention of configuration via two different mechanisms. ChemBioChem 6:1999–2004
73. Chien CH, Shann YJ, Sheu SY (1996) Site-directed mutations of the catalytic and conserved amino acids of the neuraminidase gene, nanH, of *Clostridium perfringens* ATCC 10543. Enzyme Microb Technol 19:267–276
74. Wang QP, Graham RW, Trimbur D, Warren RAJ, Withers SG (1994) Changing enzymatic-reaction mechanisms by mutagenesis - conversion of a retaining glucosidase to an inverting enzyme. J Am Chem Soc 116:11594–11595
75. Watson JN, Newstead S, Dookhun V, Taylor G, Bennet AJ (2004) Contribution of the active site aspartic acid to catalysis in the bacterial neuraminidase from *Micromonospora viridifaciens*. FEBS Lett 577:265–269
76. Guo X, Sinnott ML (1993) A kinetic-isotope-effect study of catalysis by *Vibrio cholerae* neuraminidase. Biochem J 294:653–656

Influenza Virus Sialidase and Structure-Based Drug Design

Jeffrey C. Dyason and Mark von Itzstein

Abstract Structure-based drug design was critical in the development of the currently available drugs for treating influenza virus infection. Recent reports show there are in fact two different groups of influenza virus A sialidases, with the main structural difference between the two groups being a flexible loop region in the vicinity of the active site. This report looks at how the crystal structures of the new group 1 sialidases have been used to design new influenza virus A sialidase inhibitors which would target the flexible loop region. Although there have been several studies using molecular dynamics and docking techniques, to date there has been only one report of inhibitors that have been designed, synthesised and proved to target the flexible loop region.

1 Introduction

Influenza continues to be an ongoing problem despite the existence of vaccines and drugs. Disease outbreaks can occur relatively quickly as witnessed with the recent emergence of the influenza virus A/H1N1 pandemic. Pandemics have only originated from influenza virus A strains, as there are reservoirs of the disease in other animal populations, for example, birds, swine, etc. Influenza virus has two exposed surface glycoproteins, haemagglutinin (HA) and sialidase (neuraminidase, NA [EC 3.2.1.18]), with HA being involved in the initial binding and fusion events while NA is involved in the release of virion progeny from the cell surface. The functions of these proteins have been further elaborated in an earlier chapter (see [1]). Both of these proteins recognise the terminal sialic acid residues of cell surface oligosaccharides and are essential in the lifecycle of the virus. In this chapter, we focus entirely on the structure-based design of inhibitors of influenza virus sialidase.

J.C. Dyason • M. von Itzstein (✉)
Institute for Glycomics, Gold Coast Campus Griffith University, Southport, QLD 4222, Australia
e-mail: m.vonitzstein@griffith.edu.au

M. von Itzstein (ed.), *Influenza Virus Sialidase - A Drug Discovery Target*, Milestones in Drug Therapy, DOI 10.1007/978-3-7643-8927-7_4,

Influenza virus sialidase was first purified in the 1960s and the first X-ray crystal structure of an A/N2 sialidase was determined in the early 1980s [2, 3]. Subsequently, numerous X-ray crystal structures of influenza virus sialidase from various strains of influenza virus A and B have been determined as well as a number of NA-complex structures (e.g., with α-Neu5Ac (**1**) [4–6]) and these have been previously reviewed [7]. Importantly, in 2006 structures of the clinically important A/N1 subtype were solved [8]. Moreover, as is elaborated below the structures of the A/N1 led to the notion of two distinct phylogenetic groups of influenza virus sialidase as well as a more-than-expected flexible active site architecture.

Influenza A and B virus sialidases display a sequence homology around 30% and between influenza A virus strains less than 50%. Also, it has been noted that a number of amino acid residues are highly conserved between virtually all A and B virus subtypes [9]. The relationship between these highly conserved residues was further clarified from structural analysis, from which they were found to be associated with the enzyme's active site cavity [10–13]. The active site is made up of several distinct pockets that contain eight highly conserved amino acid residues which make direct contact with a complexed α-*N*-acetylneuraminic acid residue (α-Neu5Ac, **1**). In addition, there are a further ten invariant amino acid residues in all wild-type strains of influenza virus within the vicinity of the active site that appear to contribute to the stabilisation of the architecture of the active site [10]. The mechanism of influenza virus sialidase is thoroughly discussed in an earlier chapter [14].

In complex with the enzyme (Fig. 1), the α-Neu5Ac moiety is oriented within the active site through charge–charge interactions between the moiety's C-1 carboxylate group and a triarginyl cluster [4]. The C-5 acetamido group, embedded in a pocket of the active site, makes a number of interactions with active site amino acid residues, including hydrogen bonding of the carbonyl oxygen and the N–H to residue Arg152[1] and a buried water molecule respectively. Furthermore, the C-5 acetamido moiety's methyl group also engages in hydrophobic contacts to both Trp178 and Ile222. Finally, a number of hydrogen bond networks are formed by the C-8 and C-9 hydroxyl groups of the glycerol side chain to the carboxylate oxygen of residue Glu276, while the C-4 hydroxyl group associates with the carboxylate oxygen of Glu119. Significantly, all of the amino acid residues mentioned above are fully conserved across the wild-type strains of influenza virus known to date. A detailed analysis of the interactions between α-*N*-acetylneuraminic acid (**1**) and the influenza virus sialidase active site have been previously reported [10].

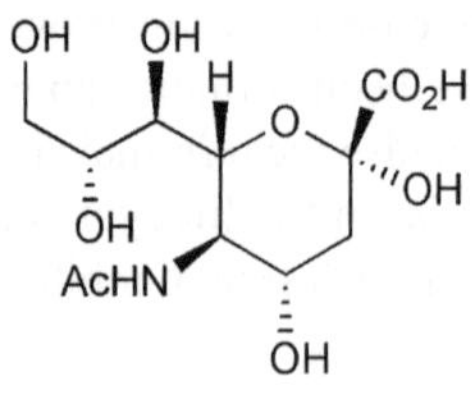

1

[1] Numbering used throughout reflects that reported for influenza A virus N2 sialidase.

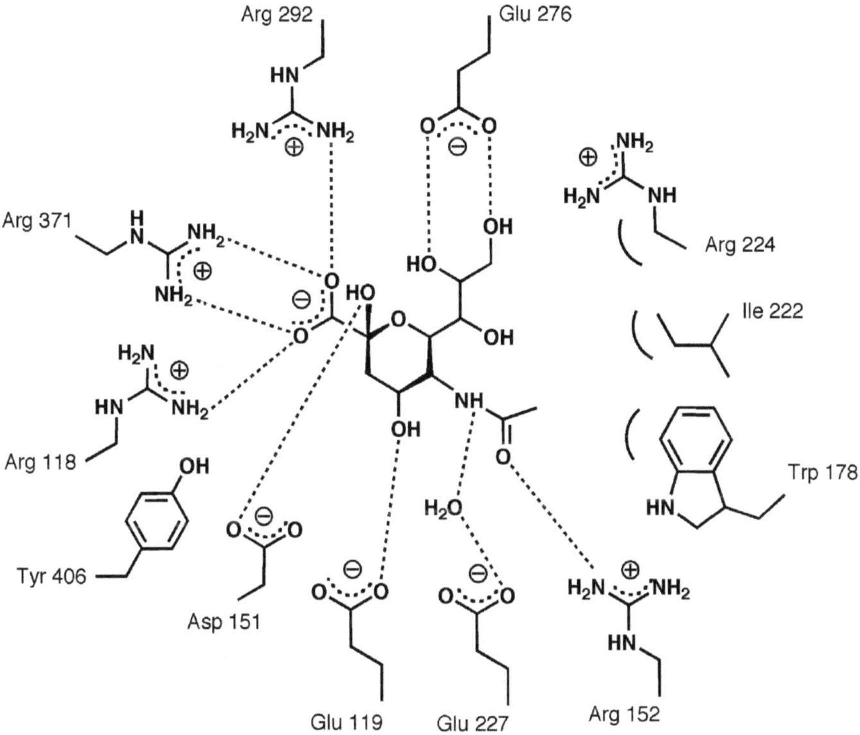

Fig. 1 Key interactions between α-*N*-acetylneuraminic acid (α-Neu5Ac, **1**) and the key influenza virus sialidase active site amino acids involved in interactions with **1**. The carboxylate of **1** forms strong electrostatic interactions with the triarginyl cluster made up of Arg 118, Arg 371 and Arg 292. Several other active site amino acids, including Asp 151, Glu 119, Glu 227, Glu 276, Arg 152, Arg 224, Trp 178, Ile 222 all have either direct hydrogen bond, water-mediated or hydrophobic interactions with functional groups associated with **1**

2 History

There are currently two drugs available for treating influenza virus infection by acting as NA inhibitors. They are Relenza© (zanamivir, **2**) marketed by Glaxo Wellcome and Tamiflu© (oseltamivir, **3**) marketed by Hoffmann–La Roche. The active form of **3** is oseltamivir carboxylate (**4**). Another NA inhibitor that was initially abandoned due to poor oral availability is peramivir (**5**). This inhibitor is now being developed by Biocryst Pharmaceuticals as an injectable drug and is currently undergoing clinical trials. A fourth promising NA inhibitor A-315675 has been under development for several years by Abbott Laboratories but has not progressed to clinical trials. A broader discussion on influenza virus drugs and inhibitors that are either available or under current development is provided in [15, 16].

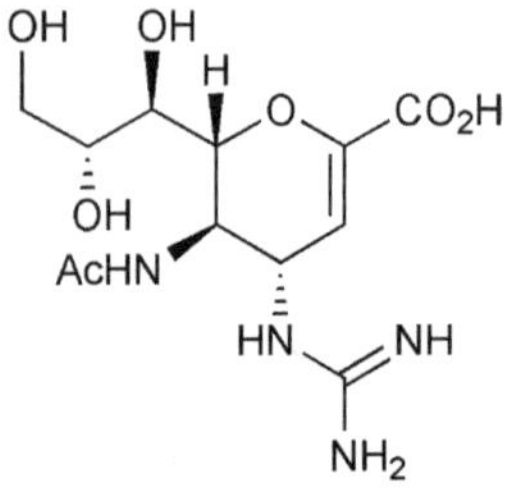

2

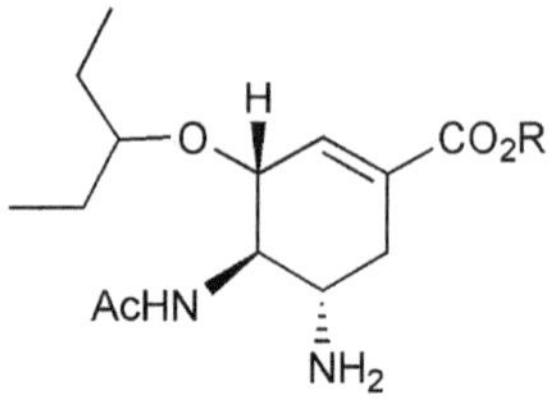

3 R = Et
4 R = H

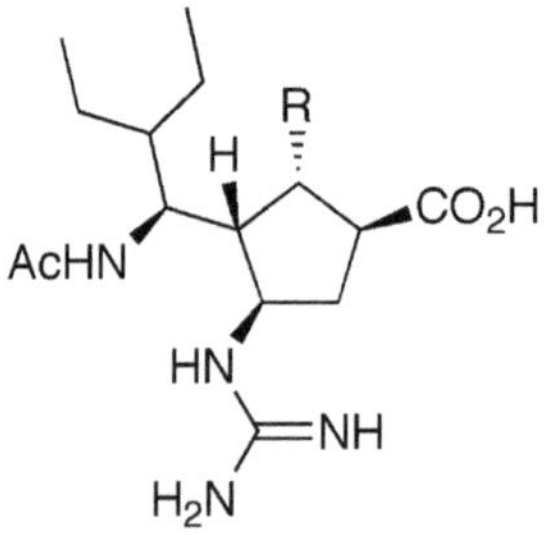

5 R = OH

All of the above NA inhibitors have been developed or optimised using the crystal structures available at the time and these included both N2 and N9 influenza A virus sialidase and several influenza B virus sialidases. The sialidase inhibitors show excellent inhibition across a wide range of influenza A and B virus strains. In 2006, Russell et al. [8] showed that for influenza virus A strains there existed two subgroups of sialidases, group 1 consisting of N1, N4, N5 and N8 sialidases and group 2 consisting of N2, N3, N6, N7 and N9 sialidases. The major structural difference between these two subgroups involved the flexibility of the ‘150-loop’ (residues 147–152) and also the reorientation of the side chain of Glu-119.

In the crystal structures of the apo group 1 sialidases (compared to the group 2 sialidase structures) the 150-loop is in an open conformation which exposes a large pocket, adjacent to the active site (Fig. 2), while the side chain of Glu-119 is reoriented and pointing in almost the opposite direction to that seen in the group 2 sialidase structures (Fig. 2). The crystal structures with several bound inhibitors of the group 1 sialidase suggest that both the 150-loop and the side chain of Glu119 adopt the more normal conformation seen in the group 2 sialidases. Under short soaking times or low oseltamivir carboxylate (**4**) concentrations however, it was possible to isolate structures where this inhibitor was bound with an open 150-loop conformation and Glu119 reoriented. This suggests that the apo group 1 sialidases in solution exist with an open 150-loop conformation and reoriented Glu119 side chain which upon binding an inhibitor or presumably the natural substrate, causes the 150-loop to close and the reorientation of the Glu119 side chain to present an active site substantially similar to that seen in group 2 sialidases. This also explains

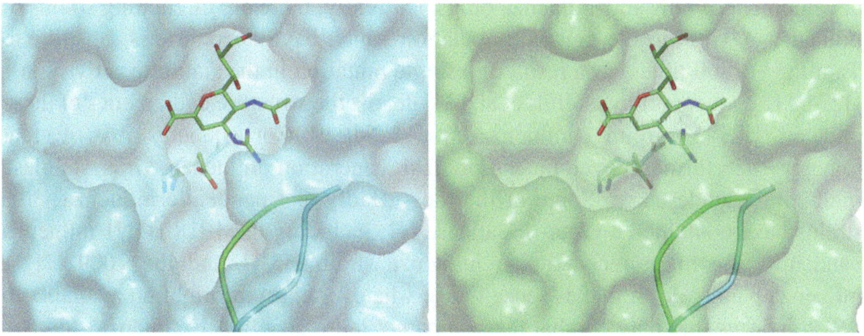

Fig. 2 The superimposed crystal structures of the 1918 N1 in the apo form (3BEQ, cyan carbons) and with zanamivir (**2**) bound (3B7E, green carbons). The *left-hand side panel* shows the surface of the apo form with the open 150-loop and zanamivir (**2**) as a reference point while the *right-hand side* shows the surface of the closed 150-loop form. Glu119 from each structure is shown as cylinders and can be seen on the floor of the binding site

why the developed sialidase inhibitors seem to be equally effective against either group 1 or group 2 sialidases.

Recently, Xu et al. [17] reported the crystal structure of the sialidase isolated from the 1918 influenza virus (H1N1) pandemic. This apo N1 sialidase showed the same active site structural features as observed in the earlier reported N1, N4 and N8 group 1 apo crystal structures, i.e. an open 150-loop and reoriented Glu119 side chain. Also, demonstrated was that on binding an inhibitor (zanamivir, **2**) the 150-loop and Glu119 side chain change conformation to resemble the active site previously seen in group 2 sialidase crystal structures. A recent crystal structure of the apo form of the swine flu pandemic (H1N1) sialidase [18] suggested that this sialidase does not have an open 150-loop. However, a more recent report shows that an inhibitor designed to lock open the 150-loop in group 1 sialidases effectively inhibits the swine flu pandemic (H1N1) sialidase [19]. This suggests that in the case of the swine flu pandemic (H1N1) sialidase the 150-loop is evenly poised between being open or closed and that a designed inhibitor could influence the loop conformation.

3 Structure-Based Drug Design (SBDD)

SBDD has played an important role in the development of influenza virus sialidase inhibitors and anti-influenza drugs [20, 21]. Aspects of the structure-based inhibitor design using the Neu5Ac2en (**6**) template are discussed elsewhere in [15]. So what influence has the existence of the most recent group 1 influenza virus sialidase structures had on the development of next generation drugs to treat influenza? As noted in the Russell et al. [8] report, the new cavity exposed by the flexible 150-loop in group 1 influenza virus sialidases suggests it may be possible to use it for the development of new anti-influenza drugs. Currently, there are circulating strains of influenza A virus in the general population which have become less sensitive to Tamiflu© (~800 times) as a result of a single mutation in the sialidase active site – H274Y [22]. A report by Collins et al. [23] details the structure and kinetics of these mutated (H274Y) group 1 influenza virus sialidases. Consequently, it is of great urgency that further development of next generation influenza virus sialidase inhibitors is undertaken. Some of the most deadly and pandemic influenza virus strains have contained a group 1 sialidase N1, including the 1918 H1N1 influenza pandemic, the more recent avian influenza strain H5N1 and the 2009 H1N1 influenza pandemic. It makes sense to use whatever extra information available to develop drugs that may combat these apparent more deadly N1 influenza strains.

4 Use of the Recent Influenza A Virus N1 Sialidase Structures in SBDD

In the intervening year since the first appearance of the open 150-loop Apo N1 structure, a number of different computational chemistry techniques have been used to study the enzyme. This includes predominantly molecular dynamics and docking experiments to study a range of sialidase properties including the flexible 150-loop, understanding the emergence of resistance, as well as how existing inhibitors interact with N1 sialidase. However, this chapter is mainly interested in structure-based drug design, so we will concentrate on the use of the recent N1 structures in the design on new influenza virus sialidase inhibitors.

4.1 Molecular Dynamics

There are several reports of molecular dynamics studies employed to investigate sialidase inhibitors [24–26]. One study by Cheng et al. [24] used a molecular dynamics approach to generate several starting protein structures that were then used to screen the National Cancer Institute (NCI) diversity set via a virtual screening protocol. In other studies by Park and Jo [25] and Wang et al. [26], molecular dynamics protocols were used to calculate binding energies of oseltamivir carboxylate (**4**) derivatives and natural products respectively. None of these reports included any assay data to verify their theoretical predictions.

4.2 Docking

A number of reports [24, 27–36] describe the use of docking techniques to design inhibitors that should make use of the open 150-loop available in N1 sialidases.

These studies predominantly employ derivatives of either zanamivir (**2**) or oseltamivir carboxylate (**4**) while large chemical databases like the NCI database were also screened. From these reports only two groups actually evaluated the predicted inhibitors in assays with An et al. [33] describing a compound from the NCI database that, while active, was not quite as potent an inhibitor of N1 sialidase as oseltamivir carboxylate (**4**) and Wen et al. [28] study producing a micromolar inhibitor of N1 sialidase based on zanamivir (**2**) substituted at the guanidino group. No data were presented for either compound that demonstrated selectivity for N1 sialidases over N2 sialidases. In principle these designer compounds should have shown markedly less inhibition, if they were actually targeting the N1 sialidase open 150-loop that does not exist in the N2 sialidase.

A very recent report from the von Itzstein group [19] outlines the use of docking and visualisation in the design of Neu5Ac2en (**6**) derivatives substituted at the C-3 position leading to compounds such as **7**. These derivatives were synthesised and displayed better inhibition against N1 sialidase compared with N2 sialidase. A crystal structure using the group 1 sialidase N8 clearly shows that the 3-substituent occupies the 150-loop cavity as it was designed to do, locking it in an open form (Fig. 3).

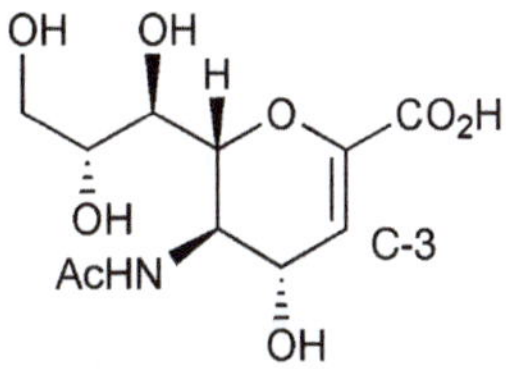

6

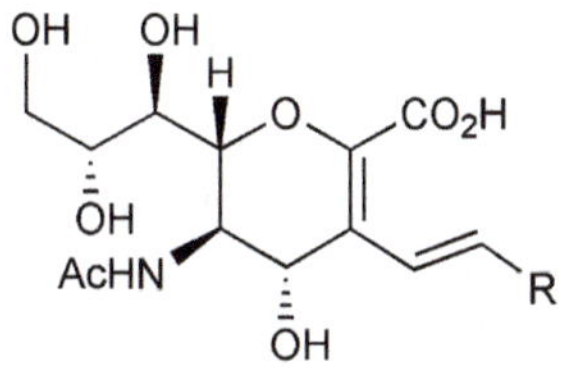

7

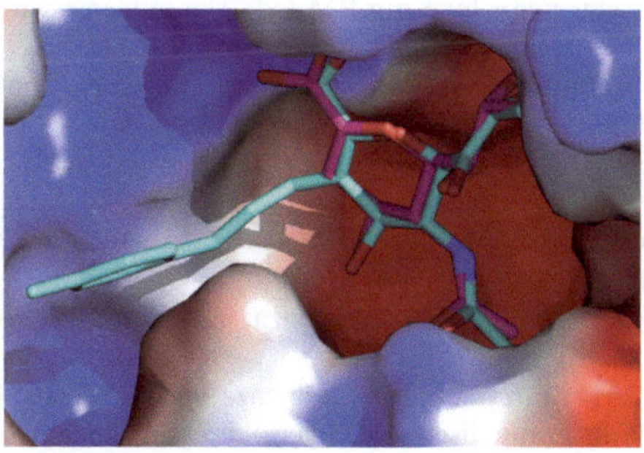

Fig. 3 Electrostatic potential map of the group 1 influenza virus sialidase N8 in complex with the C-3 functionalised Neu5Ac2en derivative **7** (cyan) where R = *p*-tolyl. The protein surface is coloured by charge (*red* = negative, *blue* = positive and *white* = neutral/hydrophobic). Superimposed on **7** is the parent template in magenta, Neu5Ac2en (**6**). The core Neu5Ac2en moiety of the C-3 functionalised derivative **7** orients in an identical fashion to the parent template **6**

5 Conclusions

The use of sialidase crystal structures in the design and development of the currently available influenza drugs is well documented. The appearance of oseltamivir-resistant strains of seasonal influenza confirms that there is a need to continue the development of next generation anti-influenza drugs and the crystal structures of the N1 sialidases have now provided such an opportunity.

References

1. Klenk HD (2011) Influenza virus virology. In: von Itzstein M (ed) Influenza virus sialidase: a drug discovery target. Springer, Heidelberg
2. Colman PM, Varghese JN, Laver WG (1983) Structure of the catalytic and antigenic sites in influenza virus neuraminidase. Nature 303:41–44
3. Varghese JN, Laver WG, Colman PM (1983) Structure of the influenza virus glycoprotein antigen neuraminidase at 2.9 A resolution. Nature 303:35–40
4. Varghese JN, McKimm-Breschkin JL, Caldwell JB, Kortt AA, Colman PM (1992) The structure of the complex between influenza virus neuraminidase and sialic acid, the viral receptor. Proteins 14:327–332
5. Bossart-Whitaker P, Carson M, Babu YS, Smith CD, Laver WG, Air GM (1993) Three-dimensional structure of influenza A N9 neuraminidase and its complex with the inhibitor 2-deoxy 2,3-dehydro-N-acetyl neuraminic acid. J Mol Biol 232:1069–1083
6. Burmeister WP, Henrissat B, Bosso C, Cusack S, Ruigrok RW (1993) Influenza B virus neuraminidase can synthesize its own inhibitor. Structure 1:19–26
7. Colman PM (2009) New antivirals and drug resistance. Annu Rev Biochem 78:95–118
8. Russell RJ, Haire LF, Stevens DJ, Collins PJ, Lin YP, Blackburn GM, Hay AJ, Gamblin SJ, Skehel JJ (2006) The structure of H5N1 avian influenza neuraminidase suggests new opportunities for drug design. Nature 443:45–49
9. Smith FI, Palese P (1989) Variation in the influenza virus genes: epidemiological, pathogenic and evolutionary consequences. In: Krug RM (ed) The influenza virus. Plenum, New York, pp 319–350
10. Colman PM (1994) Influenza virus neuraminidase: structure, antibodies, and inhibitors. Protein Sci 3:1687–1696
11. Laver G, Garman E (2002) Pandemic influenza: its origin and control. Microbes Infect 4:1309–1316
12. Colman PM (1994) Influenza virus neuraminidase: structure, antibodies, and inhibitors. Protein Sci 3:1687–1696
13. Laver G, Garman E (2002) Pandemic influenza: its origin and control. Microbes Infect 4:1309–1316
14. Chan J, Bennet AJ (2011) Enzymology of influenza virus sialidase. In: von Itzstein M (ed) Influenza virus sialidase: a drug discovery target. Springer, Heidelberg
15. Thomson R, von Itzstein M (2011) The development of carbohydrate-based influenza virus sialidase inhibitors. In: von Itzstein M (ed) Influenza virus sialidase: a drug discovery target. Springer, Heidelberg
16. Streicher H, Stanley M (2011) The development of non-carbohydrate-based influenza virus sialidase inhibitors. In: von Itzstein M (ed) Influenza virus sialidase: a drug discovery target. Springer, Heidelberg
17. Xu XJ, Zhu XY, Dwek RA, Stevens J, Wilson IA (2008) Structural characterization of the 1918 influenza virus H1N1 neuraminidase. J Virol 82:10493–10501

18. Li Q, Qi J, Zhang W, Vavricka CJ, Shi Y, Wei J, Feng E, Shen J, Chen J, Liu D et al (2010) The 2009 pandemic H1N1 neuraminidase N1 lacks the 150-cavity in its active site. Nat Struct Mol Biol 17:1266–1268
19. Rudrawar S, Dyason JC, Rameix-Welti MA, Rose FJ, Kerry PS, Russell RJ, van der Werf S, Thomson RJ, Naffakh N, von Itzstein M (2010) Novel sialic acid derivatives lock open the 150-loop of an influenza A virus group-1 sialidase. Nat Commun 1:113
20. von Itzstein M (2007) The war against influenza: discovery and development of sialidase inhibitors. Nat Rev Drug Discov 6:967–974
21. von Itzstein M, Thomson R (2009) Anti-influenza drugs: the development of sialidase inhibitors. Handb Exp Pharmacol 189:111–154
22. Gubareva LV, Webster RG, Hayden FG (2001) Comparison of the activities of zanamivir, oseltamivir, and RWJ-270201 against clinical isolates of influenza virus and neuraminidase inhibitor-resistant variants. Antimicrob Agents Chemother 45:3403–3408
23. Collins PJ, Haire LF, Lin YP, Liu JF, Russell RJ, Walker PA, Skehel JJ, Martin SR, Hay AJ, Gamblin SJ (2008) Crystal structures of oseltamivir-resistant influenza virus neuraminidase mutants. Nature 453:1258–U1261
24. Cheng LS, Amaro RE, Xu D, Li WW, Arzberger PW, McCammon JA (2008) Ensemble-based virtual screening reveals potential novel antiviral compounds for avian influenza neuraminidase. J Med Chem 51:3878–3894
25. Park JW, Jo WH (2010) Computational design of novel, high-affinity neuraminidase inhibitors for H5N1 avian influenza virus. Eur J Med Chem 45:536–541
26. Wang YT, Chan CH, Su ZY, Chen CL (2010) Homology modeling, docking, and molecular dynamics reveal HR1039 as a potent inhibitor of 2009 A(H1N1) influenza neuraminidase. Biophys Chem 147:74–80
27. Durrant JD, McCammon JA (2010) Potential drug-like inhibitors of Group 1 influenza neuraminidase identified through computer-aided drug design. Comput Biol Chem 34:97–105
28. Wen WH, Wang SY, Tsai KC, Cheng YSE, Yang AS, Fang JM, Wong CH (2010) Analogs of zanamivir with modified C4-substituents as the inhibitors against the group-1 neuraminidases of influenza viruses. Bioorg Med Chem 18:4074–4084
29. Sun JY, Cai SX, Yan N, Mei H (2010) Docking and 3D-QSAR studies of influenza neuraminidase inhibitors using three-dimensional holographic vector of atomic interaction field analysis. Eur J Med Chem 45:1008–1014
30. Chen CY, Huang HJ, Tsai FJ, Chen CYC (2010) Drug design for Influenza A virus subtype H1N1. J Taiwan Inst Chem Eng 41:8–15
31. Li Y, Zhou BC, Wang RX (2009) Rational design of Tamiflu derivatives targeting at the open conformation of neuraminidase subtype 1. J Mol Graph Model 28:203–219
32. D'Souza C, Kanyalkar M, Joshi M, Coutinho E, Srivastava S (2009) Search for novel neuraminidase inhibitors: design, synthesis and interaction of oseltamivir derivatives with model membrane using docking, NMR and DSC methods. Biochim Biophys Acta-Biomembr 1788:1740–1751
33. An JH, Lee DCW, Law AHY, Yang CLH, Poon LLM, Lau ASY, Jones SJM (2009) A novel small-molecule inhibitor of the avian influenza H5N1 virus determined through computational screening against the neuraminidase. J Med Chem 52:2667–2672
34. Rungrotmongkol T, Frecer V, De-Eknamkul W, Hannongbua S, Miertus S (2009) Design of oseltamivir analogs inhibiting neuraminidase of avian influenza virus H5N1. Antiviral Res 82:51–58
35. Mitrasinovic PM (2009) On the structure-based design of novel inhibitors of H5N1 influenza A virus neuraminidase (NA). Biophys Chem 140:35–38
36. Garcia-Sosa AT, Sild S, Maran U (2008) Design of multi-binding-site inhibitors, ligand efficiency, and consensus screening of avian influenza H5N1 wild-type neuraminidase and of the oseltamivir-resistant H274Y variant. J Chem Inf Model 48:2074–2080

The Development of Carbohydrate-Based Influenza Virus Sialidase Inhibitors

Robin Thomson and Mark von Itzstein

Abstract This chapter presents a review of the development of influenza virus sialidase inhibitors based on a carbohydrate scaffold, predominantly that of the natural 'ligand' of the enzyme, *N*-acetylneuraminic acid (Neu5Ac). These inhibitors include hydrolytically stable substrate-like compounds, product mimics, and transition-state-like compounds. The major focus of the chapter, reflecting the most intensively researched area of inhibitor development, is on the use of the dihydropyran scaffold of the 2,3-unsaturated-Neu5Ac derivative, Neu5Ac2en, a putative transition state mimic. Structure-based drug design targeting conserved residues of the sialidase active site using this template leading to the development of the potent and selective inhibitor, and anti-influenza drug, zanamivir (Relenza™), as well as subsequent developments towards next generation inhibitors, are discussed.

Abbreviations

NA	Neuraminidase (sialidase)
Neu5Ac	*N*-acetylneuraminic acid
Neu5Ac2en	2-deoxy-2,3-didehydro-*N*-acetylneuraminic acid
MDCK	Madin-Darby canine kidney (cell line)

1 Introduction

The vital role of influenza virus sialidase in the lifecycle of the virus, as discussed in the previous chapters, implied that the enzyme's activity should be a good target for therapeutic intervention against the virus. The design of anti-influenza therapies

R. Thomson (✉) • M. von Itzstein
Institute for Glycomics, Griffith University Gold Coast Campus, Southport, QLD 4222, Australia
e-mail: r.thomson@griffith.edu.au

M. von Itzstein (ed.), *Influenza Virus Sialidase - A Drug Discovery Target*,
Milestones in Drug Therapy, DOI 10.1007/978-3-7643-8927-7_5,

against this antigenic surface protein was considered a challenge, however, due to the propensity for facile and rapid mutation (antigenic drift) of the viral genome. With the determination of the X-ray crystal structure of influenza A virus N2 sialidase in the early 1980s [1], the highly conserved nature of critical catalytic and substrate-binding active site residues across all influenza A and B virus strains (discussed in [2]) was revealed, providing the opportunity for rational, structure-based drug design targeting conserved residues of the sialidase. Knowledge gained from the influenza virus sialidase X-ray crystal structures, studies of the enzyme mechanism (discussed in [3]), and structure-activity relationship (SAR) studies on sialic acid derivatives, has since been utilised in the development of a number of potent and selective influenza virus sialidase inhibitors (reviewed in [4, 5]), including the anti-influenza drugs zanamivir (Relenza™) and oseltamivir (Tamiflu™). This chapter presents a review of influenza virus sialidase inhibitors developed on carbohydrate scaffolds, in particular those based on the natural 'ligand' of the enzyme, *N*-acetyl-α-neuraminic acid.

2 Interactions of *N*-acetylneuraminic Acid with Influenza Virus Sialidase

The relative importance to binding to sialidases of the different functional groups of the natural 'ligand' *N*-acetyl-α-neuraminic acid (α-Neu5Ac, **1**) was determined initially through SAR studies using synthetic sialic acid derivatives in the 1960s and 1970s. In particular these studies identified an almost [6] absolute requirement for a free carboxylic acid at C-1 [7] and a distinct preference for the acetylamino group at C-5 [8, 9] for substrate (sialoside) binding and hydrolysis. Introducing a double bond into the pyranose ring of Neu5Ac between C-2 and C-3 produced a sialidase inhibitor [10], Neu5Ac2en (**2**). The importance of these features for binding to influenza virus sialidase was subsequently clarified and refined with the advent of X-ray crystallographic structure information of influenza virus sialidases in complex with Neu5Ac-based compounds, in the 1980s and 1990s. The binding and interactions of α-Neu5Ac (**1**) within the catalytic site are illustrated in Fig. 1, and described here in brief to assist the following discussions of the relative effectiveness of the different types of sialic acid-based influenza virus sialidase inhibitors. Detailed analysis of these interactions is provided in [2].

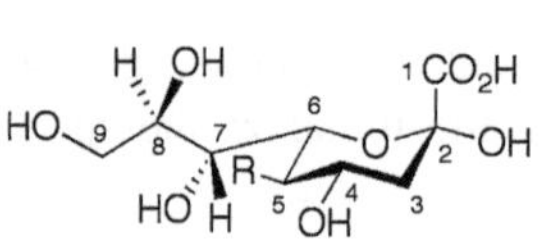

1 R = NHAc (α-Neu5Ac)

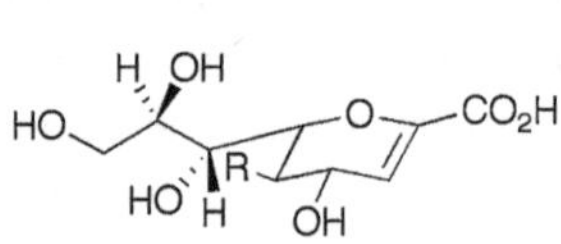

2 R = NHAc (Neu5Ac2en)

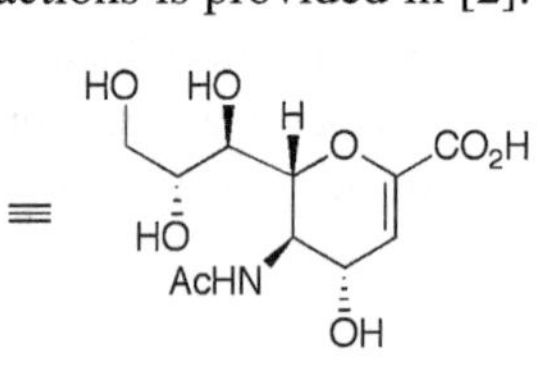

2 Neu5Ac2en

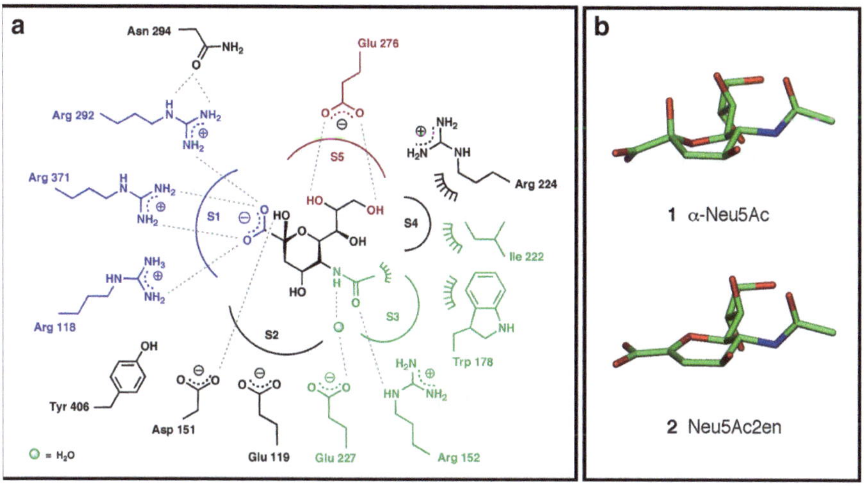

Fig. 1 (**a**) Schematic representation of the interactions of α-Neu5Ac (**1**) with conserved active site residues in the influenza A/N2 sialidase active site (PDB 2bat) [11], with interactions coloured according to the dominant functional group interactions of the C-1 carboxylate (*purple*), the C-5 acetamido group (*green*), and the C-8:C-9 diol of the glycerol side-chain (*maroon*). (**b**) *Upper panel*: A view of the α-boat form of α-Neu5Ac (**1**) (PDB 2bat). *Lower panel*: A view of the half-chair form of Neu5Ac2en (**2**) (PDB 1f8b) that results in virtually identical positioning of the substituents in the active site to those of α-Neu5Ac. In these representations, the base of the active site is below 1 (or 2); the C-2 hydroxyl group of 1 projects out of the active site into the solvent. (Figure 1b was prepared using AstexViewer [12])

In describing substrate or inhibitor binding, the active site can be effectively divided into five regions or subsites (S1–S5, Fig. 1a) [13], with subsites 1, 2, 3, and 5 binding the substituents of bound α-Neu5Ac. Subsite 4, which is primarily a hydrophobic region, is not occupied by any portion of α-Neu5Ac. While these subsites are common to influenza A and B virus sialidases, an additional cavity adjacent to subsites 1 and 2 has been observed in apo structures of a subgroup of influenza A virus sialidases (N1, 4, 5, and 8; the 'group-1' sialidases). This cavity is formed as a result of a more open conformation of a flexible protein loop, the so-called '150-loop' [14] (described in more detail in [2]). When bound into the sialidase active site, α-Neu5Ac (**1**) adopts an α-boat-like conformation [11, 15] (Fig. 1b), with the C-2 hydroxyl group of α-Neu5Ac (or the aglycon of the sialoside substrate) projecting out of the active site [11]. In this conformation, the substituents at C-2, C-4, C-5, and C-6 are held in pseudo-equatorial orientation, projecting into subsites 1, 2, 3, and 5, respectively, of the active site. Essentially identical active site interactions to those of α-Neu5Ac (**1**) are seen for the substituents on the 2,3-unsaturated derivative Neu5Ac2en (**2**) [11, 15]. No significant differences are observed in the active site amino acid side-chain conformations between the apo sialidase structures and structures with either α-Neu5Ac (**1**) or Neu5Ac2en (**2**) bound [15].

3 Carbohydrate Templates for Development of Sialidase Inhibitors

Sialic acid-based inhibitors of influenza virus sialidase have included substrate, product, and transition-state-like compounds. While certain features of synthetic sialosides could lead to hydrolytically stable compounds, some of which were inhibitory of enzyme activity, the majority of research on carbohydrate-based inhibitors has focussed on monosaccharide derivatives of Neu5Ac.

3.1 Hydrolytically Stable Substrate Analogues

Two general approaches to the development of sialidase-resistant substrate analogues (sialosides) as inhibitors of influenza virus sialidase have been used: modification of the glycosidic linkage and/or modification of the Neu5Ac residue of the sialoside. In each case the substrate analogues maintain the α-glycosidically linked Neu5Ac residue that is also recognised by influenza virus haemagglutinin providing the potential for dual action inhibitors [16].

Replacing the glycosidic oxygen with sulfur, nitrogen, or carbon, produces sialosides that are resistant to sialidase hydrolysis, in contrast to the corresponding *O*-glycosides. The extent to which these Neu5Ac sialosides are inhibitory depends upon the attached aglycon unit [17–19] (see for example Figs. 2: **3** vs. **4** [18] and **5** vs. **6** [19]). For example, sialidase-resistant thioglycoside analogs of gangliosides competitively inhibited GM_3 hydrolysis by the sialidases of different subtypes of human and animal influenza viruses down to micromolar levels, with the nature of the carbohydrate aglycon modulating activity; changing from the natural substrate α(2,6)-linked Gal (**3**) to α(2,6)-linked Glc (**4**), resulted in a 100-fold decrease in apparent K_i values for inhibition of GM_3 hydrolysis [18].

Sialidase-resistant *O*-sialosides have been prepared by incorporating modification into the sialic acid residue which renders the glycosidic bond more stable, or which alters the binding of the sialoside in the active site. In an approach aimed specifically at developing sialoside-based inhibitors of influenza virus haemagglutinin that would also be resistant to, and inhibitory of, influenza virus sialidase, a small electronegative group was introduced at C-3 of the Neu5Ac residue to stabilise [20] the glycosidic linkage [16]. Distearoylphophatidyl-ethanolamine (DSPE) *O*-sialosides with 3-axial-F (**7**) or -OH, or 3-equatorial-OH on the Neu5Ac residue, inhibited haemagglutination and hemolysis caused by a human influenza A (H3N2) virus, and were resistant to hydrolysis by influenza virus sialidases (in contrast to the parent, 3-H, sialoside **8**), however only the 3-axial-F derivative **7** was inhibitory of the enzyme activity [16]. The dual inhibitory activity of 3-ax-F-Neu5Ac-2-α-DSPE (**7**) translated to a synergistic inhibitory effect at the micromolar level (IC_{50} 5.6 μM) against viral infection in cultured MDCK cells [16].

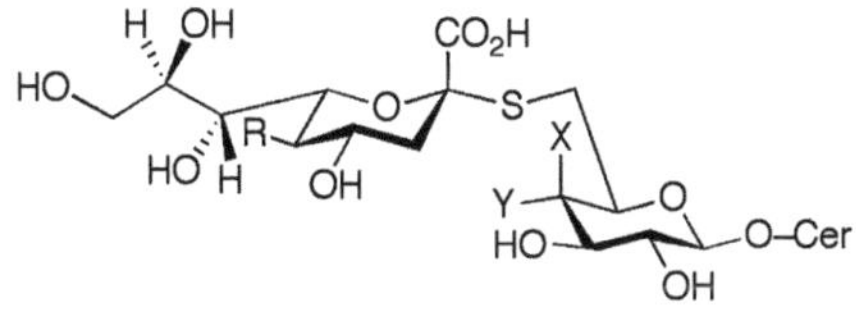

3 R = NHAc; X = OH, Y = H (Gal): K_i 100 μM
4 R = NHAc; X = H, Y = OH (Glc): K_i 2.8 μM

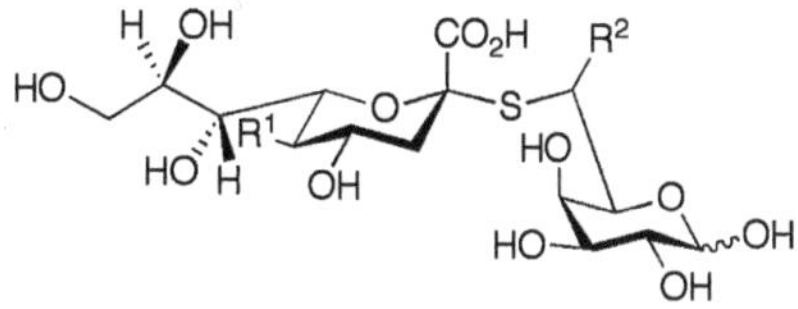

5 R^1 = NHAc; R^2 = H: K_i > 5 mM (A/N1)
6 R^1 = NHAc; R^2 = CH_3: K_i 300 μM (A/N1)

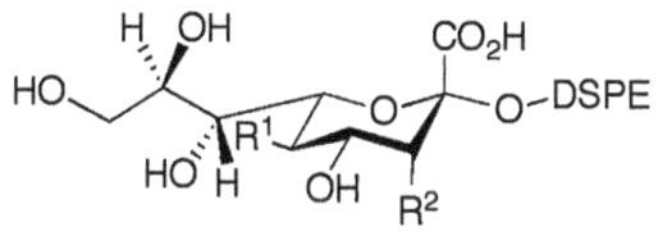

7 R^1 = NHAc; R^2 = F: IC_{50} 30 μM (A/N2)
8 R^1 = NHAc; R^2 = H: enzyme substrate

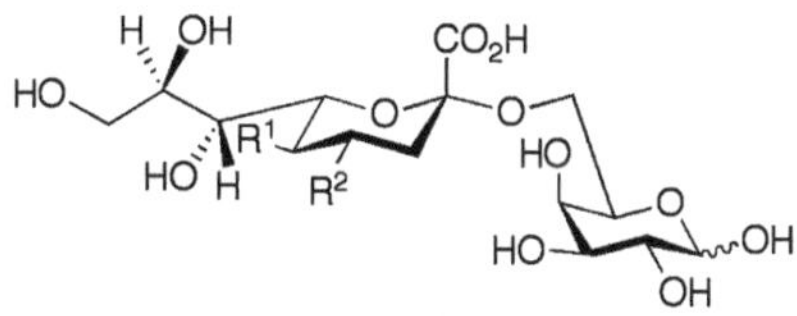

9 R = NHAc; R^2 = NH_2: K_i 150 μM (A/N1)
10 R = NHAc; R^2 = OH: enzyme substrate

O-DSPE =

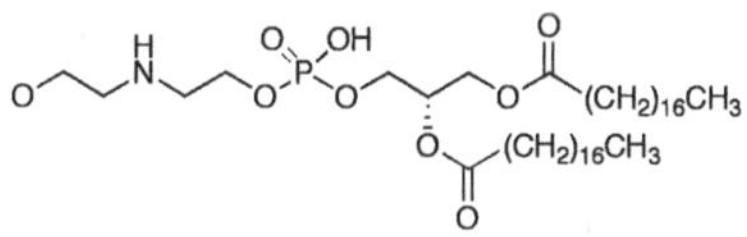

Fig. 2 Influenza sialidase-resistant *S*- (**3–6**) and *O*- (**7, 9**) glycosides

Modification of the sialic acid moiety to enhance binding within the active site has been made by replacing the C-4 hydroxyl group of the Neu5Ac residue in Neu5Ac-α(2,6)-Gal 10 with an amino group; a substitution identified to increase binding affinity during structure-based inhibitor design on the Neu5Ac2en template [21, 22] (see Sect. 4). This change produced an *O*-sialoside 9 which was not only resistant to cleavage by influenza virus NA but was also inhibitory (K_i 150 μM) of the enzyme (H1N1 virus suspension), as was a related thiosialoside [23].

3.2 Neu5Ac and Product Analogues

N-Acetylneuraminic acid (Neu5Ac), the product of enzymatic sialoside hydrolysis, inhibits influenza virus sialidase in the millimolar region [K_i 5 mM (A/N2) [24]]. However, the low proportion (~5%) [25] of the sialidase-binding α-Neu5Ac (**1**) in the solution α/β mixture of Neu5Ac may contribute to the apparently relatively weak inhibition level. Replacing the 2-α-OH of **1** with hydrogen (**11**) or with a fluorine atom (**12**), to give defined α-stereochemistry at C-2, progressively increased inhibition of the enzyme [26] (Fig. 3). Significantly, 2-deoxy-2-α-H-Neu5Ac (**11**) showed weak, but measurable, in vivo activity against influenza A (H2N2) virus infection in mice when administered intranasally (at 100 mg/kg, reduction of virus titres to 19%; at 25 mg/kg reduction to 57% of those seen in uninfected mice) [26].

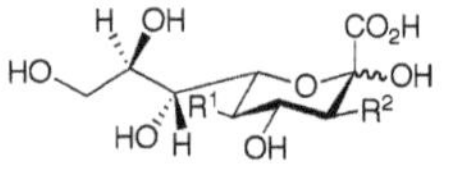

1 R^1 = NHAc; R^2 = H: K_i 5 mM (A/N2)
13 R^1 = NHAc; R^2 = F: app K_i 8 μM (A/N1)

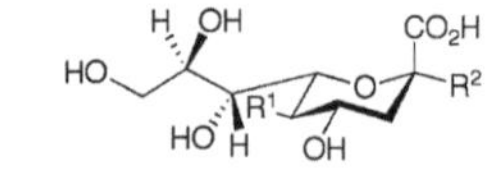

11 R^1 = NHAc; R^2 = H: K_i 300 μM (A/N2)
12 R^1 = NHAc; R^2 = F: K_i 40 μM (A/N2)

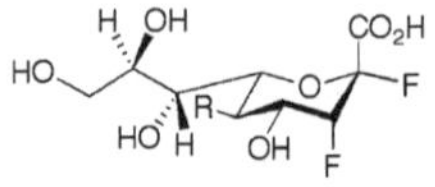

14 R = NHAc

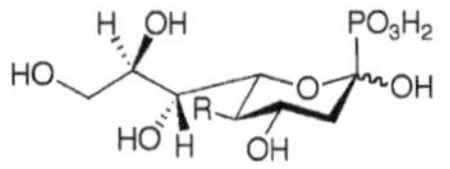

15 R = NHAc; K_i 200 μM (A/N2)
16 R = OH; K_i 200 μM (A/N2)

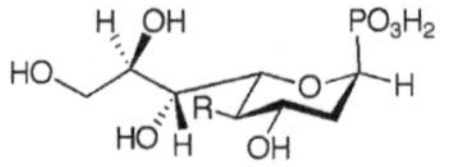

17 R = NHAc (aPANA)
Inhibition comparable to Neu5Ac

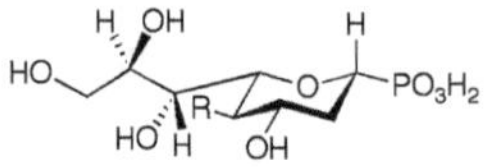

18 R = NHAc (ePANA)
Inhibition ~100-fold stronger than Neu5Ac

Fig. 3 Influenza virus sialidase inhibitors (**1, 11–13, 15–18**) based on substituted Neu5Ac

Investigating the position adjacent to the anomeric centre, replacement of the C-3 equatorial hydrogen with an electronegative fluorine atom, giving 3-eq-F-Neu5Ac (**13**), produced a competitive inhibitor of bacterial and viral sialidases: influenza A/N1 sialidase was inhibited with an apparent K_i of 8 μM [27]. Introducing fluorine at both C-2 and C-3 of Neu5Ac (e.g. 2-α-F:3-ax-F, **14**) has been used in the development of a novel class of mechanism-based enzyme inactivators [28, 29], to probe the catalytic mechanism of sialidases. The anomeric fluoride acts as a good leaving group, while the fluorine atom adjacent to the anomeric centre inductively destabilises the formation of a positive charge during the enzyme transition state [20]. This destabilisation of positive charge results in a slowing of both the formation and hydrolysis of a covalent sialosyl-enzyme intermediate, and can be used to effectively ‘trap’ this intermediate species [20, 28]. While 2,3-difluorinated-Neu5Ac has been used to trap covalent enzyme intermediates for trypanosomal ‘regular’ [20, 29] and ‘trans’ [28] sialidases, at the time of writing corresponding studies with influenza virus sialidase have not yet been reported.

The free carboxylate of α-Neu5Ac (**1**) contributes significantly to the binding energy of the ligand in the active site through interaction with three highly conserved arginine residues [30] (Fig. 1a). Two studies have examined the effect on enzyme inhibition of ‘strengthening’ this charge–charge interaction by replacing the monobasic carboxylate with a dibasic phosphonate. Replacing the carboxylate group of Neu5Ac with a phosphonate (**15**) resulted in an apparently stronger inhibitor of influenza virus sialidase (IC_{50} 200 μM). Significantly, in the sialic acid analogues with the phosphonate group, the natural preference for an acetamide group at C-5 was over-ridden, with the 5-hydroxy analogue **16** having equivalent inhibition to the 5-acetamido derivative **15** [31].

A study of phosphonate analogues of 2-deoxy-2-H-Neu5Ac highlighted the energetic advantage of an inhibitor that binds to the active site in its solution conformation. 2-Deoxy-2-α-H phosphonate (**17**, aPANA), with stereochemistry at C-2 (axial acid) equivalent to α-Neu5Ac (**1**), showed sialidase inhibition ‘equivalent to Neu5Ac’, while the C-2 isomer [2-deoxy-2-β-H phosphonate (**18**, ePANA)] was 100-fold more inhibitory against A/N2 sialidase [32]. X-ray crystal structures of A/N9 complexes of the two isomers showed that a PANA 17 bound in an α-boat-like

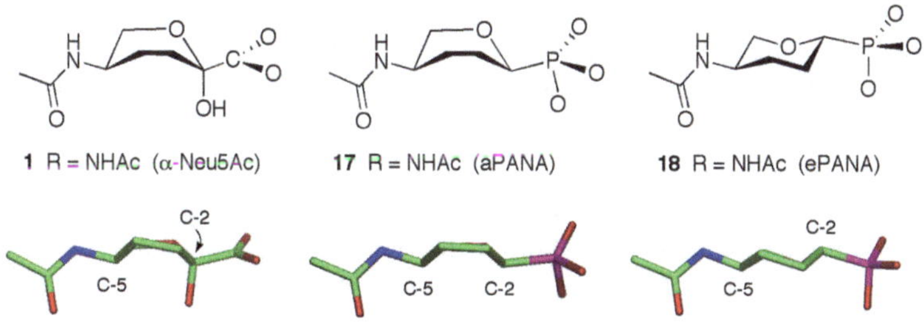

Fig. 4 Comparison of the structures of α-Neu5Ac (**1**), 2-deoxy-2-α-H-phosphonate (**17**, aPANA) and 2-deoxy-2-β-H-phosphonate (**18** ePANA) in complex with influenza virus A/N2 sialidase (PDB: 2bat [11], 1inw [32], and 1inx [32] respectively) [the 4-OH and glycerol side-chain have been omitted for clarity], showing the different conformations of the pyranose ring that position the acidic group and the C-4 OH, C-5 NHAc, and C-6 glycerol side chain for optimum interaction with active site residues. In these representations, the base of the active site is above 1 (or the PANA analogues)

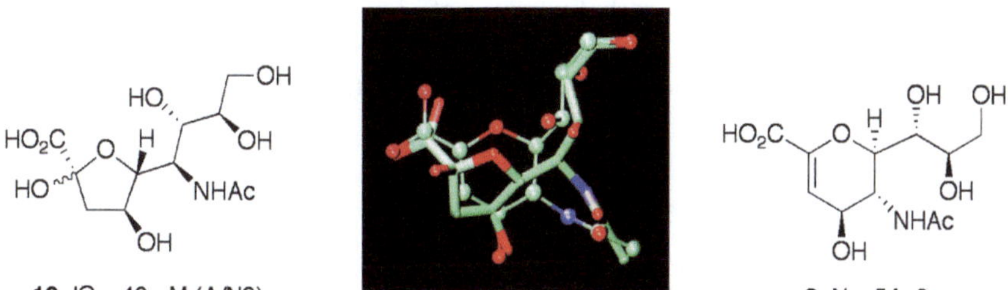

19 IC_{50} 40 μM (A/N9)
Inhibition comparable to Neu5Ac2en **2**

2 Neu5Ac2en

Fig. 5 Comparison of the structures of Neu5Ac furanoside isomer **19** [33] and Neu5Ac2en (**2**). The superimposition [34] of the structures of **19** (cylinder) and **2** (ball and stick) from their complexes with A/N9 sialidase shows the closely equivalent positioning of the main functional groups on the furanose and dihydropyran rings, respectively, despite the different positions of the central ring. In these representations, the base of the active site is below the inhibitors. (Central figure is reproduced, with permission, from [34])

conformation (as seen for α-Neu5Ac), while e-PANA 18 bound in its solution $^2C_5$ conformation [32]. Despite the conformational difference in the central ring (Fig. 4), there were no major positional differences of the phosphonate or other functional groups in **17** and **18** [32], which were essentially as seen in the α-Neu5Ac (**1**):A/N2 complex [11]. The similarity of the functional group interactions suggested that the chair to boat conformational change upon binding to the active site contributed to the lower inhibitory activity of aPANA (**17**) [32].

Moving away from the pyranose scaffold of the natural substrate/product of sialidase, in exploration of an alternative template for inhibitor design Shiba and coworkers [33] prepared furanose isomers of Neu5Ac, with isomer **19** showing inhibition of influenza A sialidase comparable to Neu5Ac2en (**2**) (85% cf 95%

inhibition respectively at 1 μM [33]; IC_{50} 40 μM against A/N9 sialidase [34]). Subsequently, comparison of the X-ray crystal structures of **19** and the inhibitor Neu5Ac2en (**2**) in complex with A/N9 sialidase (Fig. 5) [34] showed that despite the position of the furanose ring of **19** being significantly displaced from that of Neu5Ac2en (**2**), the carboxylic acid, glycerol side-chain, C-4 hydroxyl, and C-5 acetamido groups, of both inhibitors occupied the same relative positions in the active site and had similar interactions with the enzyme [34]. This finding was the beginning of the development by Biocryst Pharmaceuticals of the potent cyclopentane-based influenza sialidase inhibitor, and current investigational drug, peramivir [5, 34] (see following chapter).

The studies described above provide a number of potential templates for inhibitor design, in particular the 2-deoxy-2-β-H phosphonate (ePANA, **18**) and furanoside **19** analogues of Neu5Ac. Both **18** and **19** highlight the 'scaffolding' nature of the central carbohydrate ring – which provides appropriate positioning of the substituents to interact with the different binding pockets of the sialidase active site. This concept has formed the basis of the development of potent influenza virus sialidase inhibitors on non-carbohydrate scaffolds (discussed in [35]). The most potent inhibitors described to date on a carbohydrate template, however, are those based on the dihydropyran scaffold of Neu5Ac2en (**2**).

3.3 Neu5Ac2en and Derivatives: Mimics of the Putative Sialidase Sialosyl Cation Transition State Intermediate

The 2,3-unsaturated *N*-acetylneuraminic acid derivative, 2-deoxy-2,3-didehydro-*N*-acetylneuraminic acid (Neu5Ac2en, **2**), was first chemically synthesised in the mid-1960s [36], and was found to be a competitive inhibitor of bacterial and viral sialidases [10, 36]. Neu5Ac2en was subsequently reported to be a component of normal serum, urine, and saliva [37]. Neu5Ac2en is a micromolar inhibitor of influenza virus sialidase [IC_{50} 19 μM [10]; K_i 4 μM (A/N2) [38]], but also inhibits a number of other viral, bacterial, and mammalian sialidases to similar levels [10, 38]. Interestingly, with this template, the methyl ester was also a competitive inhibitor of influenza virus sialidase, though 1,000-fold weaker than the free acid [6].

Inhibition of sialidases by Neu5Ac2en is generally 100–1,000-fold stronger than that of the product of sialoside hydrolysis, Neu5Ac. This inhibitory potency, compared to the K_m values for substrate hydrolysis (e.g. 10^{-3} [39] to 10^{-4} M [24]), taken together with the half-chair conformation of the dihydropyran ring [40], suggested that the unsaturated derivative was a mimic of the putative sialosyl cation transition state **20** (see [3]) of the enzyme reaction [39]. X-ray crystallographic studies of Neu5Ac2en in complex with influenza virus [11, 15] and other sialidases subsequently confirmed that Neu5Ac2en bound in its solution-state half-chair conformation.

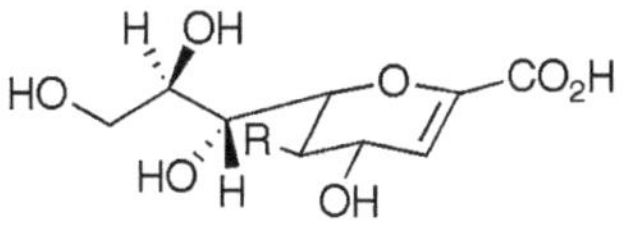

2 R = NHC(O)CH_3 K_i 4 μM (A/N2)

21 R = NHC(O)CF_3 K_i 0.8 μM (A/N1)

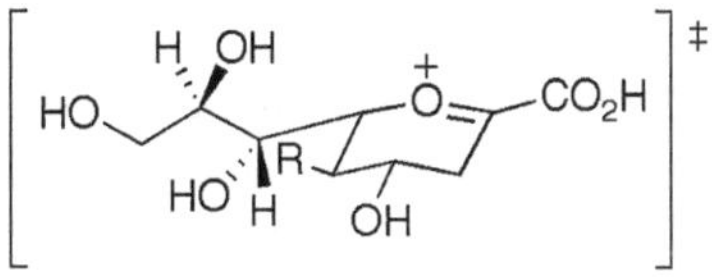

20 R = NHAc

Detailed modifications (epimerization and deoxygenation of the glycerol side-chain [41, 42], modification at C-4 [6, 43], at C-5 [10, 44], and at C-1 and C-5 [45]) of Neu5Ac2en were made during the 1980s and early 1990s, but the inhibitory effects of these compounds were reported in the main only against the most frequently examined 'test case' sialidase, bacterial *V. cholerae* sialidase [45, 46], with relatively few compounds assessed for inhibition of influenza virus sialidases [6, 10, 47]. Of these, from a range of C-5 acylamino derivatives, the *N*-trifluoroacetyl derivative (Neu5TFAc2en, **21**) gave the strongest inhibition seen so far (influenza A/Melb, IC_{50} 5 μM cf Neu5Ac2en (**2**) IC_{50} 19 μM [10]; K_i 0.79 μM [10]).

Neu5TFAc2en 21 inhibited viral replication in cell culture with the mechanism of action determined to be the inhibition of release of sialic acids from the viral envelope, resulting in the aggregation of newly synthesised virus particles at the infected cell surface, giving evidence of one of the roles of sialidase in viral replication [48]. However, the inhibitory activity of Neu5TFAc2en in vitro did not translate to inhibition in a mouse model of influenza virus infection [49]. This was postulated due to either metabolism of **21** [49] or to its rapid clearance, as later demonstrated for the parent template, Neu5Ac2en [50].

4 Structure-Based Inhibitor Design on the Neu5Ac2en Template: The Development of the Anti-influenza Drug, Zanamivir (Relenza™)

Up until the mid-1980s, variations to the structure of Neu5Ac2en had not improved the activity against any sialidase beyond that of the C-5 fluoroacetamido derivatives [10] for viral or bacterial sialidases. Furthermore, Neu5Ac2en (**2**) and Neu5TFAc2en (**21**) inhibited to a similar extent sialidases of viral or other origin [10, 38]. Neu5Ac2en, however, was a particularly useful template in that its half-chair conformation [40] appeared to mimic the putative enzyme transition state and did not require a conformational change on binding, and there were a range of possibilities for functional group modification. What was required was a way to increase potency and promote selectivity for inhibition of the viral enzyme.

The availability of X-ray crystal structure data of influenza A N2 sialidase [1, 51, 52] in the early 1980s, and subsequently of complexes with α-Neu5Ac (**1**) [11] and Neu5Ac2en (**2**) [11], provided the opportunity for rational, structure-based design of inhibitors. 'Structure-directed' modifications were anticipated to enhance binding affinity, and to generate compounds with greater specificity for influenza virus compared to mammalian sialidases. The highly conserved nature of the amino acids that positioned the sialic acid derivatives in the active site [52], despite high antigenic variability of the viral sialidase, not only meant that sialidase was a good drug target against the highly mutating virus, but also provided the opportunity to develop compounds that could target all influenza A and B viruses. In addition, by targeting inhibitor interactions to conserved residues it was hoped that the generation of viable mutant viruses would not be facile.

The design of more effective inhibitors of influenza virus sialidase was approached through detailed analysis of the X-ray crystal structure of influenza A/N2 sialidase using computer-assisted manual examination and molecular modelling studies [21, 22, 30]. One aspect of the design process was the use of the predictive software program GRID [53], to highlight potential energetically favourable interactions between various functional group probes (such as carboxylates, amines, methyl groups, and phosphates) and residues within the A/N2 NA active site [21]. Overlaying the functional group 'hot spots' with the position of Neu5Ac2en (**2**) in the A/N2:Neu5Ac2en complex identified potential substitutions on the Neu5Ac2en template that would be predicted to increase binding affinity.

Interestingly the GRID study showed no predicted interaction between the protein and a hydroxyl group probe in the area of the C-4 hydroxyl group of Neu5Ac2en (subsite 2, Fig. 1a), at an energy level that showed potential interactions for the glycerol side-chain hydroxyl groups [21]. This apparent lack of significant interaction is reflected in the comparable inhibitory potency of Neu5Ac2en (**2**) and its 4-deoxy analogue (**2**: K_i 40 μM; 4-deoxy-Neu5Ac2en: K_i 80 μM [54]). The fact that the C-4 position of Neu5Ac2en (**2**) is somewhat 'under-utilised' in normal binding of the inhibitor into the sialidase active site highlighted this position on the template as useful for increasing interaction with the protein while maintaining the unchanged important interactions of the other functional groups.

Significant among the predicted favourable interactions, was that for a protonated amino group in the predominantly negatively charged subsite adjacent to the C-4 hydroxyl group of Neu5Ac2en (subsite 2, Fig. 1a). From this finding it was anticipated that replacement of the C-4 hydroxyl group of Neu5Ac2en (**2**) by an amino group should enhance binding affinity as a result of salt bridge formation between the C-4 amine and conserved glutamic acid residue Glu119 in this region [21, 22]. Importantly, it was evident that subsite 2 may also be able to accommodate larger, more basic functional groups than the amine. This analysis, together with chemical intuition, led to the conclusion that a guanidino group incorporated at C-4 of Neu5Ac2en, although bulky and potentially occupying most of subsite 2, could form substantial interactions with two highly conserved acidic amino acid residues (Glu119 and Glu227) via its terminal nitrogens.

Keeping in place the carboxylate, the C-5 acetamido group, and the glycerol side-chain, all of which have significant interactions with the sialidase active site [11], modification at C-4 was carried out to introduce the amino and guanidino functionalities. The two target C-4 substituted Neu5Ac2en derivatives, 4-amino-4-deoxy-Neu5Ac2en (**22**) and 4-deoxy-4-guanidino-Neu5Ac2en (**23**), were synthesised [55] and found to provide significantly improved inhibition of sialidases from both influenza A and B viruses [22, 38] compared to the parent inhibitor Neu5Ac2en (**2**).

4.1 Biological and Structural Evaluation of 4-Amino- (22) and 4-Guanidino- (23) 4-Deoxy-Neu5Ac2en

Both 4-amino- (**22**) and 4-guanidino- (**23**) 4-deoxy-Neu5Ac2en were found to be competitive inhibitors of influenza A and B viral sialidases in an in vitro fluorimetric enzyme activity assay. While introduction of the amino group at C-4 (**22**) increased inhibitory potency by 10–100-fold compared to the parent inhibitor Neu5Ac2en 2, the 4-guanidino derivative **23** was significantly more potent with K_i values 1,000-fold lower than Neu5Ac2en [22, 38] (Fig. 6). In addition, 4-guanidino-Neu5Ac2en (**23**) exhibits slow binding kinetics [22, 58] further decreasing the apparent K_i, e.g. against A/N2 sialidase from 10^{-9} to 10^{-10} M [22]. Extensive studies with 4-guanidino-Neu5Ac2en (**23**) determined it to be a potent (nM), broad spectrum inhibitor of the sialidase activity of medically relevant human influenza A (H1N1, H2N2, and H3N2) and B viruses [22, 59, 60], as well as of all nine influenza A sialidase subtypes [61].

4-Amino- (**22**) and 4-guanidino- (**23**) 4-deoxy-Neu5Ac2en efficiently prevented in vitro and in vivo replication of both influenza A and B viruses [22] (reviewed in [62–65]). The 4-guanidino derivative **23** was particularly active, inhibiting replication of laboratory-passaged human influenza A and B viruses in tissue culture with nanomolar IC_{50} values [22, 66]. As anticipated, the sialidase inhibitor was found to inhibit the release of new virus particles from infected cells, with cell-associated virus particles found in the presence of drug. No cytotoxicity was observed

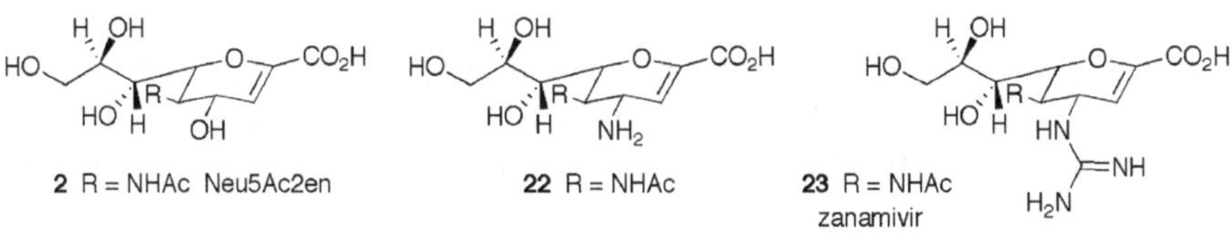

2 R = NHAc Neu5Ac2en | **22** R = NHAc | **23** R = NHAc zanamivir

2	**22**	**23**
K_i 4 μM (A/N2)	K_i 40 nM (A/N2)	K_i 1 nM to 0.2 nM (A/N2)
K_i 20 μM (B)	K_i 1 μM (B)	K_i 10 nM to 0.7 nM (B)
Human	Human	Human
K_i 12 μM (Neu1)	K_i 9.3 mM (Neu1)	K_i 1 mM (Neu1)
K_i 14 μM (Neu2)		K_i 13 μM (Neu2)

Fig. 6 Data for inhibition of influenza virus [22, 38] and human [22, 56, 57] sialidase activity in vitro by Neu5Ac2en (**2**), 4-amino-4-deoxy-Neu5Ac2en (**22**), 4-deoxy-4-guanidino-Neu5Ac2en (**23**)

at concentrations up to 10 mmol/L where cytotoxicity was observed for the other anti-influenza drugs [66]. In mouse and ferret models of influenza, intranasally administered **23** was found to be 100–1,000 times more active than the existing anti-influenza drugs amantadine and ribavirin, respectively [22].

4.1.1 Structural Analysis

X-ray crystallographic studies of the 4-amino- (**22**) [22] and 4-guanidino- (**23**) [22, 67] 4-deoxy-Neu5Ac2en derivatives in complex with influenza A sialidases showed that the compounds bound essentially in the manner predicted, with the basic amino and guanidino groups interacting with one, or both, of the glutamic acid residues in the C-4 binding domain, validating the structure-based design of the inhibitors [22]. In occupying the C-4 binding domain, the bulkier guanidino substituent of **23** displaces a water molecule, which would account for some entropic gain, as well as providing a logical reason for the reported slow binding properties of the compound [22]. Importantly, with regard to the potential for generation of drug resistant mutants, there was virtually no change in the orientation of amino acid side-chains within the catalytic domain in complexes of **22** [22] and **23** [22, 67] with influenza A virus sialidase.

4.1.2 Selectivity

In contrast to the parent structure Neu5Ac2en (**2**), 4-amino- (**22**), and in particular 4-guanidino- 23, 4-deoxy-Neu5Ac2en show high selectivity (**23** 10^3–10^6 fold) for inhibition of influenza A and B viral sialidases, over sialidases from mammalian, and bacterial, sources [22, 38]. Recent studies with recombinant forms of the four human sialidases characterised to date, confirmed millimolar inhibition of 4-guanidino derivative **23** against Neu1 (lysosomal sialidase) [56], but revealed micromolar inhibition of Neu3 (plasma membrane) [56] and Neu2 (cytosolic) [56, 57] sialidases (K_i values 4 μM and 13 μM, respectively [56]) by **23**. However, this inhibition is still substantially ($\geq$ 1,000-fold) weaker than the low nanomolar concentrations at which **23** blocks the activity of influenza virus sialidase. In addition, and significant with regard to potential inhibition of the human sialidases, is the finding that radiolabelled **23** does not enter either uninfected or influenza virus-infected MDCK cells [68].

The inhibitory specificity of **22**, and in particular **23**, for influenza virus over bacterial and mammalian sialidases has been rationalised [38, 67] on the basis of the active site architecture of the different sialidases. Bacterial sialidases appear not to have the appropriate shape of, or amino acids in, the C-4 binding domain that would permit tight binding of the C-4 guanidinyl group of **23** [67]. An X-ray crystal structure of 4-guanidino-Neu5Ac2en (**23**) in complex with recombinant human Neu2 (cytosolic) sialidase, however, shows interaction between a (potentially conserved [69]) glutamic acid residue in the C-4 binding domain and the guanidino group of **23** [57], which may provide a rationale for the micromolar level of inhibition of Neu2 by **23**.

4.1.3 The Development of Zanamivir (Relenza™)

4-Deoxy-4-guanidino-Neu5Ac2en (**23**) was licensed by Glaxo in 1990 as a lead drug candidate for the treatment of influenza virus infection under the generic name zanamivir, and gained regulatory approval as the first sialidase-targeting anti-influenza drug in 1999, under the tradename Relenza. Zanamivir is administered via oral inhalation, which delivers the drug directly to the epithelium of the lower and upper respiratory tract, the primary sites of influenza infection [63]. Intravenously administered zanamivir (which is still in the investigational phase) also shows efficacy in treatment of influenza A infection [70–72]. The clinical efficacy of zanamivir is addressed in [73].

5 Further Developments of Carbohydrate-Based Influenza Virus Sialidase Inhibitors

5.1 Towards Second-Generation Zanamivir: The Development of Long-Acting Sialidase Inhibitors

The polar nature of zanamivir limits its oral bioavailability [74], and it was found to be inactive against influenza virus infection in mice when delivered by this route. This was not unexpected given the known rapid excretion of orally administered sialic acid derivatives including the parent inhibitor Neu5Ac2en (**2**) [50]. Further developments on the 4-deoxy-4-guanidino-Neu5Ac2en (zanamivir) template have therefore been principally directed towards producing derivatives with improved pharmacokinetic properties. In this context, a prodrug form **24** of zanamivir, in which the carboxylate group is esterified with a long alkoxyalkyl chain [ester = $-(CH_2)_2O(CH_2)_{17}CH_3$] chosen to counteract the high hydrophilicity of the molecule, was reported to show significant protective effects against influenza (H1N1) infection in mice upon oral or intraperitoneal administration [75].

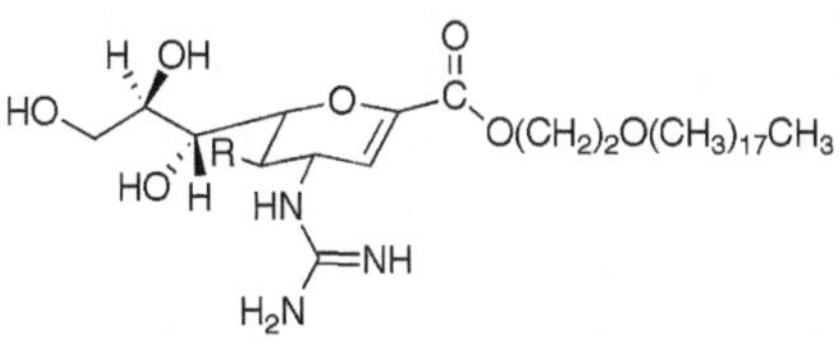

24 R = NHAc

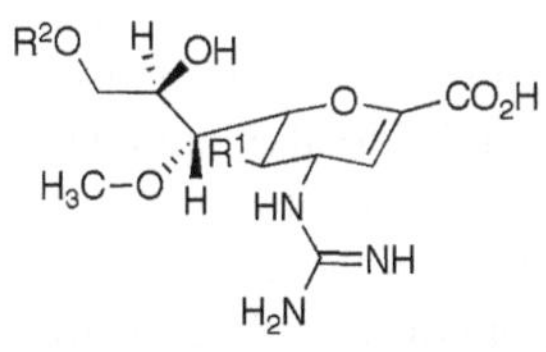

25 R^1 = NHAc, R^2 = H (R-125489)

26 R^1 = NHAc, R^2 = $C(O)C_7H_{15}$ (R-118958, CS8958)

Another, and significant, approach used to reduce the hydrophilicity of zanamivir is a modification of the glycerol side-chain. In this respect, a valuable observation was that the one functional group on zanamivir **23** that makes no direct interactions with the influenza virus sialidase protein is the C-7 hydroxyl group which points out of the active site and is exposed to bulk solvent [30, 67]. This fact has been explored and exploited in the development of C-7 functionalised zanamivir derivatives, both in terms of improving the pharmacokinetic profile and in the development of multivalent zanamivir constructs. Importantly, while the glycerol side-chain is altered, the C-7 functionalised zanamivir derivatives still retain the significant interactions of the C-8 and C-9 hydroxyl groups with Glu276.

Modifications at the C-7 position of zanamivir, such as replacement of the C-7 hydroxyl group with hydrogen, fluorine, methoxy, ethoxy, or azide, were found to have minimal or no adverse effect on inhibitory activity against influenza A (N1) sialidase or viral replication in cell culture [76]. From these derivatives, the 7-*O*-methyl ether **25** (R-125489, laninamivir), which showed inhibitory potency similar to zanamivir in sialidase enzyme assay and in cell culture [76–78] and in a mouse model of influenza infection [78], has advanced as a potential second-generation form of zanamivir. Preclinical studies in mice showed that a more lipophilic C-9 octanoyl ester [79] prodrug form of **26** (R-118958, CS-8958, laninamivir octanoate), administered intranasally, was rapidly converted by esterases to the active metabolite laninamivir, which was retained in the lungs with an elimination half-life of 41.4 h [80]. Significantly, administration of a single dose of laninamivir octanoate (post-infection) had similar efficacy to zanamivir administered twice daily [81], while a strong protective effect was also observed when the drug was administered as a single dose up to 7 days prior to infection [77, 78, 81, 82], which is more effective than zanamivir in the same situation [77, 78, 81]. This prolonged efficacy seen in mice, potentially related to the observed increased residence time of laninamivir in the lungs [80], has also been observed in human trials, with the active compound detected in urine of healthy human subjects for more than six days after a single inhalation of drug [77]. In contrast, zanamivir was not detected beyond 18 h after inhalation [77]. R-118958 X [Daiichi Sankyo and Biota] has progressed through phase III clinical trials as a long-acting, inhaled, influenza virus sialidase inhibitor, for once only treatment of, and once weekly prophylactic protection from, influenza virus infection.

5.1.1 Multivalent Conjugates

The significant level of enzyme inhibition and in vitro antiviral activity retained when zanamivir is functionalised at C-7 with sterically small groups (described above), extended to alkylation [83] or acylation (carbamate derivatives) [84] of the C-7 hydroxyl group of zanamivir with relatively long hydrophobic chains. This has been exploited in the development of polymer-supported multivalent [85, 86] and more discrete di- [87, 88], tri- [85], and tetravalent zanamivir conjugates [85, 89] [reviewed in [90]], in particular using a carbamate linkage at C-7 (as in **27**, Fig. 7).

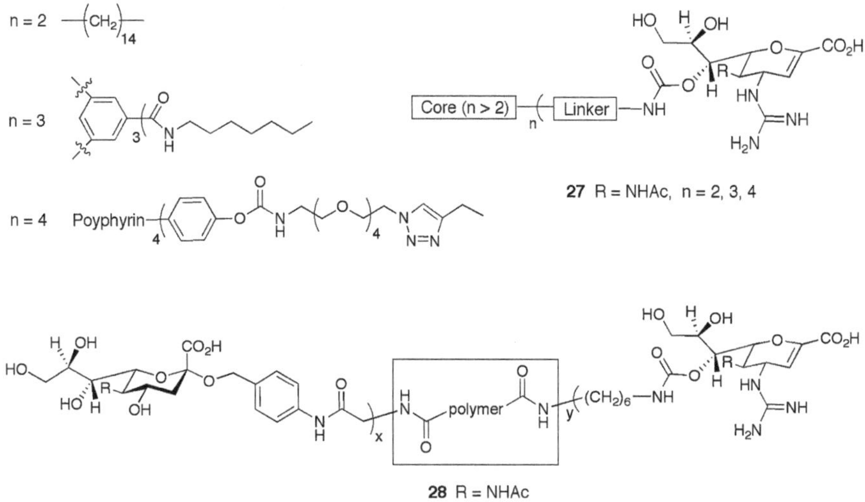

27 R = NHAc, n = 2, 3, 4

28 R = NHAc

Fig. 7 Examples of multivalent zanamivir (**27**) [85, 87, 89] and mixed zanamivir:α-sialoside (**28**) [91] conjugates

These compounds explore the potential for increased binding affinity to the tetrameric sialidase, or cross-linking of sialidase proteins.

Although showing slightly weaker sialidase inhibitory activity than zanamivir itself, the most effective multimeric compounds (see **27**, $n = 2$, 3, or 4 for examples) have significantly greater (10–100-fold) relative inhibitory potency, after correction for valency, than monomeric zanamivir in cell-based viral replication assays [85–87]. This potent in vitro activity translates to effective protection in vivo in rodent models of influenza virus infection upon intranasal administration of the conjugate [85–88], at a lower drug level than for monovalent zanamivir [85, 87]. The ability to cross-link and aggregate virus particles, visualised by electron microscopy for a divalent conjugate [87], may contribute to this inhibitory potency.

Importantly, the multivalent zanamivir conjugates have similar long-acting efficacy (e.g. after administration of a single dose seven days prior to infection [85, 87]) in rodents to that seen with 7-*O*-methyl-zanamivir, laninamivir (**25**) (described above), that can be related to long residency times in the lungs (100-fold greater concentration than zanamivir at 168 h) [87, 88]. With their favourable characteristics of potent activity and prolonged efficacy, multivalent zanamivir conjugates such as **27** ($n = 2$) are viewed as potential candidates for once a week, low dose, therapeutic agents for the prevention or treatment of influenza virus infection [87]; the dimeric forms have been forwarded for development [Biota and Daiichi Sankyo] as long-acting, inhaled sialidase inhibitors (under the class name Flunet®).

The concept of multivalent zanamivir conjugates has recently been developed further to incorporate additional aspects of antiviral activity. In a tetrameric zanamivir conjugate on a porphyrin core/scaffold **27** ($n = 4$), the porphyrin core was shown to act as a 'photosensitiser' that led to inactivation of virus particles under laboratory light in the presence of the conjugate [89]. It was surmised that this inactivation, which was not seen for the porphyrin alone and was disrupted in the presence of excess free zanamivir, was attributable to viral inactivation by singlet oxygen due to sensitisation of the porphyrin moiety which is brought into close proximity with the virus upon binding of the conjugated zanamivir to the sialidase active site [89].

In an alternative presentation of multivalent zanamivir, a polymer (**28**, Fig. 7) displaying both zanamivir (attached via a C-7 carbamate) and Neu5Ac (attached as an α-glycoside) residues provided enhanced antiviral efficacy in vitro (influenza A virus growth in MDCK cells) compared to either mono- or polyvalent zanamivir or Neu5Ac, or a mixture of both homo-polymers [91]. The hetero-polymer has the advantages that zanamivir protects the Neu5Ac *O*-sialoside from cleavage by viral sialidase, and the strong interaction of zanamivir with the sialidase should hold the Neu5Ac sialoside in close proximity to the virus for potential interaction with haemagglutinin of the same virus. The synergistic antiviral effect of the hetero-polymer is potentially due to blocking two events in the viral replication cycle – cell binding through the haemagglutinin and virus release facilitated by the sialidase [91].

5.2 *Further Developments on the Neu5Ac2en Template and Mimetics*

In work towards the development of potent and selective influenza virus sialidase inhibitors, significant efforts have been dedicated to the manipulation of each position of the Neu5Ac2en template, in particular during, and since, the development of zanamivir (**23**) (reviewed in [4, 92, 93]). In essentially all inhibitors on the carbohydrate dihydropyran template, as well as on non-carbohydrate templates (see [35]), a carboxylic acid is maintained as the acidic group [with the exception of phosphonic acid derivatives of Neu5Ac2en [45] – evaluated against *V. cholerae* NA – and on the cyclohexene template [35, 94–96] and an acetamido group is at C-5. The 'opposing' carboxylate and acetamido groups form one binding axis, to subsites 1 and 3 (Fig. 1a), of the inhibitor in the sialidase active site. The substituents that bind the orthogonal 'axis', subsites 2 and 4/5, taking the place of the C-4 hydroxyl group and glycerol side-chain of Neu5Ac, respectively, have been the most widely varied and optimised.

On the Neu5Ac2en template, with the glycerol side-chain in place, the guanidino group at C-4 appears to be optimal, both sterically and electronically, for potent sialidase inhibition. Further substitution on the terminal N^3 amine of the guanidino group with a hydroxyl or amino group **29** reduced inhibitory activity to a level comparable to Neu5Ac2en (**2**) [97], suggesting disruption of the extensive H-bonding network observed between the guanidino group of **23** and the active site residues [22, 67]. Extending the linker between C-4 and the substituent amino group, produced inhibitory activity significantly weaker than Neu5Ac2en [98]; in the case of the 4-*O*-amidinomethyl derivative **30**, in addition to steric considerations, the significant interactions [67] of the internal (N^1) nitrogen of the guanidino group of **23** are also lost.

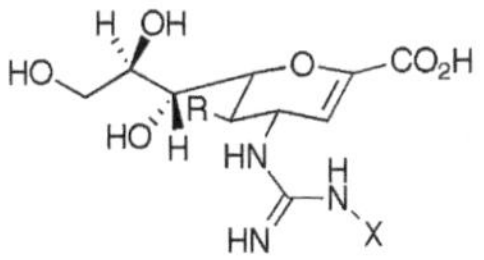

29 R = NHAc, X = OH, NH_2

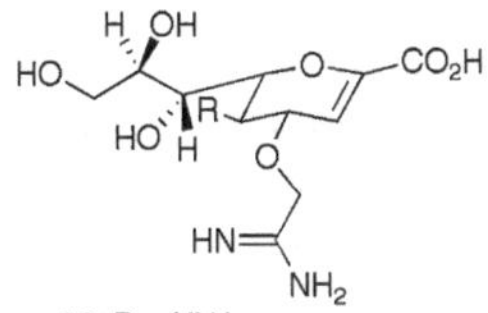

30 R = NHAc

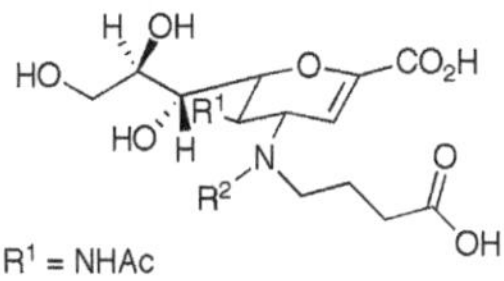

R^1 = NHAc

31 R^2 = H: IC_{50} ~70 μM (A/N1)

32 R^2 = C(=NH)NH_2: IC_{50} 7.1 μM (A/N1)

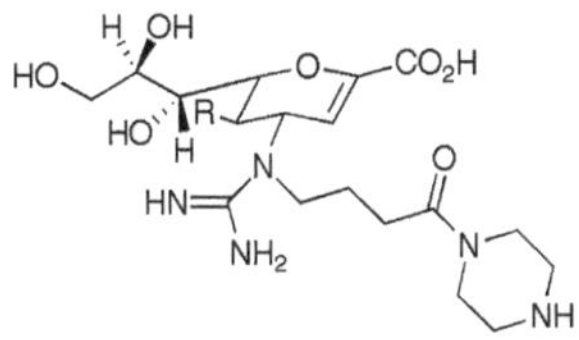

33 R = NHAc: IC_{50} 2.2 μM (A/N1)

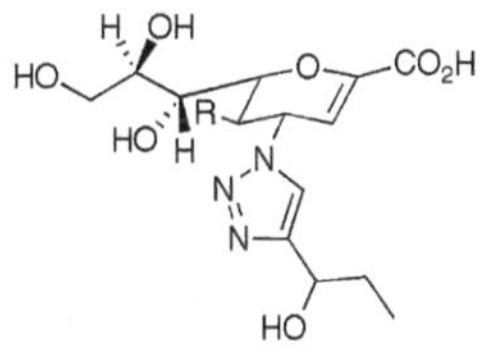

34 R = NHAc

5.2.1 Modification of 4-Amino/Guanidino-Neu5Ac2en: Towards Inhibitors Interacting with the Extended Active Site of Influenza A Group-1 Sialidases

Despite the reduced inhibitory activity seen for Neu5Ac2en derivatives with substitution on the 4-amino and 4-guanidino groups compared to the parent inhibitors **22** and **23**, interest in these compounds has been renewed with attempts to develop compounds to bind the open 150-loop form of influenza A virus group-1 sialidases where there is an additional cavity adjacent to subsites 1 and 2 [14]. A series of derivatives (e.g. **31–33**) alkylated with sterically demanding groups, in particular on the internal nitrogen of the C-4 guanidino group of zanamivir (e.g. 32, 33), showed similar inhibition to Neu5Ac2en (**2**) [99]; this still significant inhibition from a compound with such a sterically demanding group at C-4 suggested that rather than binding the closed 150-loop form of the enzyme with restricted space in subsite 2, the bulky C-4 side-chain binds the open-150-loop form of the enzyme. In a similar vein, C-4 triazole derivatives have been reported to provide varying levels of protective effect to MDCK cells against avian influenza A virus (H5N1) infection, with the 2-hydroxypropyl derivative **34** showing a protective effect approaching that of zanamivir (**23**) (EC_{50} values: **34** 6.4 microM; zanamivir (**23**) 2.8 microM) in a neutral-red uptake assay [100]. At the time of writing, however, the binding mode of these inhibitors, or their selective inhibition of group-1 sialidases had not yet been reported.

5.2.2 Replacing the Glycerol Side-Chain of 4-Amino/Guanidino-Neu5Ac2en

In SAR studies of Neu5Ac2en-based inhibitors during work towards development of second-generation drugs with improved pharmacokinetic properties, replacing the hydrophilic glycerol side-chain has been an area of particular interest. The glycerol side-chain C-8 and C-9 hydroxyl groups of sialic acid and derivatives, however, make a major contribution to anchoring sialic acid-based inhibitors in the sialidase active site through their bidentate interaction with Glu276 in subsite 5 (Fig. 1a). Without this interaction, in side-chain truncated 4-amino or 4-guanidino substituted derivatives inhibitory potency is reduced 1,000-fold [101], a significant loss in binding affinity which must be recouped by any alternative side-chain.

One approach to modulating the hydrophilicity of the glycerol side-chain of zanamivir, while maintaining the important 8,9-dihydroxy interaction with the sialidase, has been the replacement of the trihydroxy side-chain by hydroxy-substituted cyclic ethers (such as in **35**) in which the C-7 oxygen atom forms part of the ring [102]. In the most potent derivative 35, with activity comparable to zanamivir (**23**) against influenza A virus in both enzyme and plaque reduction assays, the cyclic ester diol was of the appropriate stereochemistry to enable a bidentate interaction with Glu276 [102], as in zanamivir [67]. Significantly, although the hydrophilicity of **35** might be expected to be only slightly altered compared to zanamivir, it was reported to have in vivo efficacy by oral administration in a mouse model of influenza A (N1) virus infection [102].

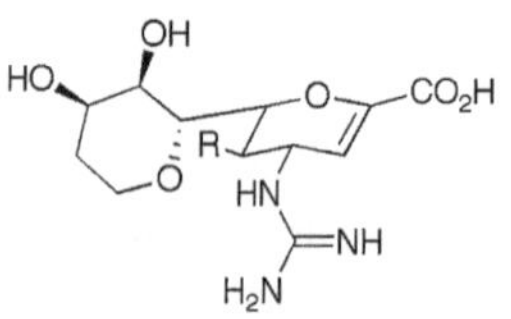

35 R = NHAc

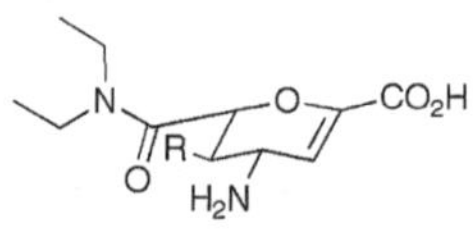

36 R = NHAc

IC_{50} 3 nM (A/N2)

IC_{50} 360 nM (B)

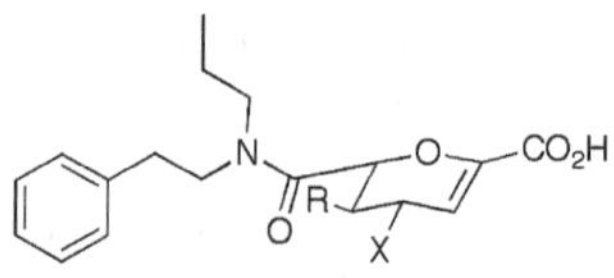

37 R = NHAc

X = H, OH, NH_2, $NHC(NH)NH_2$

IC_{50} 3 to 7 nM (A/N2)

IC_{50} 1 to 51 μM (B)

In a more radical change, researchers at Glaxo completely replaced the glycerol side-chain on the 4-amino- and 4-guanidino templates with *N*-alkylated carboxamides, accessed from the C-7 carboxylic acid obtained through oxidative cleavage of the glycerol side-chain [103–105]. Unexpectedly, at that time, the introduction of optimally sized lipophilic groups on the carboxamide produced compounds (e.g. **36**, **37**) with inhibitory activity comparable to zanamivir (**23**) against influenza A sialidase, and against influenza A virus replication both in cell culture [103, 104] and in a mouse model of influenza A infection when delivered intranasally [104]. In some cases, the strong binding interactions of the carboxamide side-chains also over-rode the influence of the C-4 substituent (as in **37**) [105]. A less favourable characteristic of the carboxamides, however, was substantially weaker ($\geq$ 100-fold weaker [104]) inhibition of influenza B virus sialidase activity, and replication in cell culture, compared to that against influenza A virus [103, 104].

The interactions of the lipophilic carboxamide side-chains within the sialidase active site that led to high binding affinity, and the weaker activity against influenza B sialidase, were identified through X-ray crystallographic studies. Complexes of the 4-amino *N*-propyl-*N*-(2-phenylethyl) carboxamide (**37**) (X = NH_2) with influenza virus A/N9 and B sialidases showed that Glu276 could reorientate to form a salt bridge with the guanidino side-chain of Arg224, creating a lipophilic pocket (bordered by the methylenes of Glu276 and Ala246) within subsite 5 (Fig. 1a) [103, 106], in which one arm of the carboxamide (e.g. propyl in **37**) was bound. The 2-phenylethyl substituent of carboxamide (**37**) bound along an extended lipophilic cleft formed between Ile222 and Ala246 (subsite 4, Fig. 1a) at the enzyme surface [103]. While the conformational change in Glu276 had only a minor effect on the positions of surrounding residues in influenza A/N9 sialidase, significant distortion of the protein was observed in type B sialidase [103, 106], providing an explanation for the difference in inhibitory potencies of the carboxamides against influenza A and B sialidases.

The strong binding affinity achieved with lipophilic side-chains at C-6 in the dihydropyran carboxamides [103], mirrored in the development of the potent inhibitor oseltamivir carboxylate (**46**, Fig. 8) with a 3-pentyl ether side-chain on the cyclohexene scaffold [108, 109], inspired further inhibitor development using

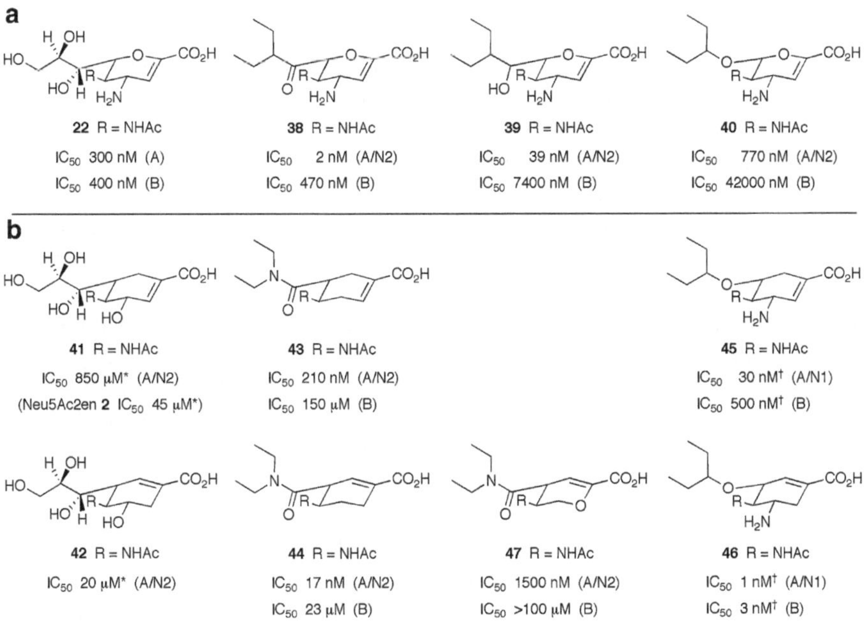

Fig. 8 Comparison of inhibitors containing hydrophobic side-chains on: (**a**) the dihydropyran scaffold; and (**b**) carbocyclic mimetic scaffolds. (*) same assay system [107]. (†) comparative inhibition values reproduced from [108]

this paradigm on both carbohydrate and non-carbohydrate (see [35]) scaffolds. In the carboxamide series (e.g. **36**) and on the cyclohexene scaffold **46**, 'diethyl' side-chains produced potent inhibitory activity. The isosteric C-7 ketone variant **38** [110] of the 4-amino *N,N*-diethyl carboxamide (**36**), showed comparable activity to the carboxamide against sialidases from both influenza A and B strains; however, the corresponding C-7 alcohol **39** was 10-fold weaker [110].

Moving from a C–C linkage at C-6 on the dihydropyran ring, to a structurally more flexible ether linkage (equivalent to the ether linkage on the cyclohexene template of oseltamivir, **46**), gave inhibitors (e.g. **40** [110]) with activity against influenza A sialidase similar to the corresponding 4-substituted Neu5Ac2en derivatives [110–112]. The weaker activity of the C-6-ether 4-amino derivative **40** compared to other dihydropyran-based compounds with a similar 'diethyl' side-chain (for example **36, 38, 39**) and to the cyclohexene-based inhibitor **46** [109] was presumed to be due to the energy penalty associated with changing from the solution-state half-chair conformation of **40** where the pendant substituents are pseudo-axial [110, 113], to the sialidase-bound Neu5Ac2en-like half-chair conformation in which the substituents are equatorially oriented.

5.2.3 Investigating the Position of the Double Bond: Moving Towards a Carbocyclic Scaffold

Molecular modelling analysis of the influenza virus sialidase conjugate with α-Neu5Ac (**1**) [30] indicated that the dihydropyran ring oxygen contributes only marginally to ligand binding, highlighting the potential for replacement of the sialic acid dihydropyran ring with an alternative scaffold. Replacing the ring oxygen of 4-amino- (**22**) and 4-guanidino- (**23**) 4-deoxy-Neu5Ac2en with sulfur gave analogues as bioactive as the parent compounds [114]. Replacement of the ring oxygen with carbon provided the opportunity to place the double bond to more closely mimic its position in the putative sialosyl cation transition state **20** of the enzyme reaction. In carbocyclic analogues of Neu5Ac2en, **41** and **42**, and of the '4-deoxy' *N,N*-diethyl carboxamides, **43** and **44**, as with cyclohexene-based **45** and **46** [108], positioning the double bond to more closely mimic the transition state indeed resulted in increased inhibitory potency (greater than an order of magnitude) against influenza A virus sialidase compared to the Neu5Ac2en double-bond configuration (Fig. 8). An attempt to incorporate the alternative double-bond position into a carbohydrate template in the form of the 4-deoxy carboxamide **47**, however, did not capture the potent activity of the corresponding cyclohexene-based carboxamide **44** [115].

In reflecting on the subtle 'push–pull' effects of binding substituents with different binding affinities on alternative scaffolds, into the enzyme active site, it is interesting to note the differences on the Neu5Ac2en scaffold. In the case of the carbocyclic analogues of Neu5Ac2en, while the closer transition state mimic, carba-Neu5Ac2en isomer **42**, indeed exhibited 40-fold stronger inhibition than the Neu5Ac2en configured analogue **41** against influenza A/N2 sialidase [107]

(Fig. 8), there is only a twofold difference in inhibition between **42** and Neu5Ac2en (**2**) itself (42 20 μM; 2 42 μM [107]). This may suggest, as might be expected, that the dihydropyran scaffold of Neu5Ac2en (**2**) is close to optimal for placement of the natural sialic acid functional groups to interact with the sialidase.

6 Summary and Future Prospects

Work towards developing effective anti-influenza therapeutics (reviewed in [116–119]) has targeted each of the viral surface proteins – M2 ion channel, haemagglutinin, and sialidase – and viral replication (e.g. polymerase inhibitors [120]). A recent novel approach being trialed is to use sialidase itself (in the form of a bacterial sialidase fusion protein [121, 122]) as a pre-emptive, reversible, strike on the receptors of the respiratory tract used by influenza virus to infect airway epithelial cells. It is inhibition of the viral sialidase, however, that has successfully provided the two major current anti-influenza drugs: Relenza™ based on a derivative of the natural carbohydrate substrate of the enzyme, *N*-acetylneuraminic acid, and Tamiflu™ built on a cyclohexene scaffold (discussed in the following chapter). The development of these potent and selective inhibitors against the rapidly mutating viral enzyme was possible through a combination of SAR studies, an understanding of the enzyme mechanism, and, importantly, the use of X-ray crystal structure data showing the binding modes of substrate and inhibitors with highly conserved active site residues. Subsequent developments towards 'second-generation' zanamivir have been based on modification of the drug's hydrophilicity, and multivalent display of the inhibitor, both of which have resulted in potential drugs with long-acting efficacy. Future advances in structure-based inhibitor design may come from targeting the 150-cavity adjacent to the active site seen in the open 150-loop form of influenza A virus group-1 sialidases (as discussed in [2]), including the N1 sialidases of the pathogenic 1918 and 2009 human pandemic H1N1 viruses and avian H5N1 virus.

Acknowledgements The financial support of the Australian Research Council (ARC) through the award of an Australian Federation Fellowship (MvI), the National Health and Medical Research Council (NHMRC), and the Honda Foundation (Australia) is gratefully acknowledged.

References

1. Colman PM, Varghese JN, Laver WG (1983) Structure of the catalytic and antigenic sites in influenza virus neuraminidase. Nature 303:41–44
2. Dyason JC, von Itzstein M (2011) Influenza virus sialidase and structure-based drug design. In: von Itzstein M (ed) Influenza virus sialidase: a drug discovery target. Springer, Heidelberg
3. Chan J, Bennet AJ (2011) Enzymology of influenza virus sialidase. In: von Itzstein M (ed) Influenza virus sialidase: a drug discovery target. Springer, Heidelberg

4. Rich JR, Gehle D, von Itzstein M (2007) Design and synthesis of sialidase inhibitors for influenza virus infections. In: Kamerling JP, Boons G-J, Lee Y, Suzuki A, Tanigichi N, Voragen AGJ (eds) Comprehensive glycoscience. Elsevier, Oxford, pp 885–922
5. Chand P (2005) Recent advances in the discovery and synthesis of neuraminidase inhibitors. Expert Opin Ther Patents 15:1009–1025
6. Flashner M, Kessler J, Tanenbaum SW (1983) The interaction of substrate-related ketals with bacterial and viral neuraminidases. Arch Biochem Biophys 221:188–196
7. Brossmer R, Holmquist L (1971) Specificity of neuraminidase. Synthesis of 5-*N*-acetyl-D-nonulosamine (5-acetamido-3,5-dideoxy-D-*glycero*-beta-D-*galacto*-nonulose) and 5-*N*-acetyl-D-neuraminamide (5-acetamido-3,5-dideoxy-D-*glycero*-β-D-*galacto*-nonulosonamide) and their methyl beta- and benzyl alpha-ketosides. Hoppe-Seyler's Z Physiol Chem 352:1715–1719
8. Meindl P, Tuppy H (1966) Darstellung und enzymatische Spaltbarkeit von a-Ketosiden der *N*-Propionyl-, *N*-Butyryl- und *N*-Benzoyl-D-neuraminsäure. Monatsh Chem 97:1628–1647
9. Brossmer R, Nebelin E (1969) Synthesis of *N*-formyl- and *N*-succinyl-D-neuraminic acid. On the specificity of neuraminidase. FEBS Lett 4:335–336
10. Meindl P, Bodo G, Palese P, Schulman J, Tuppy H (1974) Inhibition of neuraminidase activity by derivatives of 2-deoxy-2,3-dehydro-*N*-acetylneuraminic acid. Virology 58:457–463
11. Varghese JN, McKimm-Breschkin JL, Caldwell JB, Kortt AA, Colman PM (1992) The structure of the complex between influenza virus neuraminidase and sialic acid, the viral receptor. Proteins 14:327–332
12. Hartshorn MJ (2002) AstexViewer: a visualisation aid for structure-based drug design. J Comput Aided Mol Des 16:871–881
13. Stoll V, Stewart KD, Maring CJ, Muchmore S, Giranda V, Gu Y-GY, Wang G, Chen Y, Sun M, Zhao C et al (2003) Influenza neuraminidase inhibitors: structure-based design of a novel inhibitor series. Biochemistry 42:718–727
14. Russell RJ, Haire LF, Stevens DJ, Collins PJ, Lin YP, Blackburn GM, Hay AJ, Gamblin SJ, Skehel JJ (2006) The structure of H5N1 avian influenza neuraminidase suggests new opportunities for drug design. Nature 443:45–49
15. Burmeister WP, Henrissat B, Bosso C, Cusack S, Ruigrok RW (1993) Influenza B virus neuraminidase can synthesize its own inhibitor. Structure 1:19–26
16. Guo C-T, Sun X-L, Kanie O, Shortridge KF, Suzuki T, Miyamoto D, Hidari KIPJ, Wong C-H, Suzuki Y (2002) An *O*-glycoside of sialic acid derivative that inhibits both hemagglutinin and sialidase activities of influenza viruses. Glycobiology 12:183–190
17. Khorlin AY, Privalova IM, Zakstelskaya LY, Molibog EV, Evstigneeva NA (1970) Synthetic inhibitors of *Vibrio cholerae* neuraminidase and neuraminidases of some influenza virus strains. FEBS Lett 8:17–19
18. Suzuki Y, Sato K, Kiso M, Hasegawa A (1990) New ganglioside analogs that inhibit influenza virus sialidase. Glycoconjugate J 7:349–356
19. Sabesan S, Neira S, Davidson F, Duus JØ, Bock K (1994) Synthesis and enzymatic, and NMR studies of novel sialoside probes: unprecedented, selective neuraminidase hydrolysis of and inhibition by C-6-(Methyl)-Gal sialosides. J Am Chem Soc 116:1616–1634
20. Watts AG, Oppezzo P, Withers SG, Alzari PM, Buschiazzo A (2006) Structural and kinetic analysis of two covalent sialosyl-enzyme intermediates on *Trypanosoma rangeli* sialidase. J Biol Chem 281:4149–4155
21. von Itzstein M, Dyason JC, Oliver SW, White HF, Wu W-Y, Kok GB, Pegg MS (1996) A study of the active site of influenza virus sialidase: an approach to the rational design of novel anti-influenza drugs. J Med Chem 39:388–391
22. von Itzstein M, Wu WY, Kok GB, Pegg MS, Dyason JC, Jin B, Van Phan T, Smythe ML, White HF, Oliver SW et al (1993) Rational design of potent sialidase-based inhibitors of influenza virus replication. Nature 363:418–423
23. Sabesan S, Neira S, Wasserman Z (1995) Structural and functional group tuning in the design of neuraminidase inhibitors. Carbohydr Res 267:239–261

24. Walop JN, Boschman TAC, Jacobs J (1960) Affinity of *N*-acetylneuraminic acid for influenza virus neuraminidase. Biochim Biophys Acta 44:185–186
25. Friebolin H, Supp M, Brossmer R, Keilich G, Ziegler D (1980) ^{1}H-NMR investigations on the mutarotation of *N*-acetyl-*O*-neuraminic acid. Angew Chem Int Ed Engl 19:208–209
26. Colman PM, von Itzstein LM, Varghese JN, Wu W-Y, Van Phan T, White HF [Biota Scientific Management Pty. Ltd., Australia] (1992) Anti-viral compounds that bind the active site of influenza neuraminidase. *WO 92/06691*. (CAN 117:131501)
27. Hagiwara T, Kijima-Suda I, Ido T, Ohrui H, Tomita K (1994) Inhibition of bacterial and viral sialidases by 3-fluoro-*N*-acetylneuraminic acid. Carbohydr Res 263:167–172
28. Watts AG, Damager I, Amaya ML, Buschiazzo A, Alzari P, Frasch AC, Withers SG (2003) *Trypanosoma cruzi trans*-sialidase operates through a covalent sialyl-enzyme intermediate: tyrosine is the catalytic nucleophile. J Am Chem Soc 125:7532–7533
29. Watts AG, Withers SG (2004) The synthesis of some mechanistic probes for sialic acid processing enzymes and the labeling of a sialidase from *Trypanosoma rangeli*. Can J Chem 82:1581–1588
30. Taylor NR, von Itzstein M (1994) Molecular modeling studies on ligand binding to sialidase from influenza virus and the mechanism of catalysis. J Med Chem 37:616–624
31. Chan T-H, Xin Y-C, von Itzstein M (1997) Synthesis of phosphonic acid analogs of sialic acids (Neu5Ac and KDN) as potential sialidase inhibitors. J Org Chem 62:3500–3504
32. White CL, Janakiraman MN, Laver WG, Philippon C, Vasella A, Air GM, Luo M (1995) A sialic acid-derived phosphonate analog inhibits different strains of influenza virus neuraminidase with different efficiencies. J Mol Biol 245:623–634
33. Yamamoto T, Kumazawa H, Inami K, Teshima T, Shiba T (1992) Syntheses of sialic acid isomers with inhibitory activity against neuraminidase. Tetrahedron Lett 33:5791–5794
34. Babu YS, Chand P, Bantia S, Kotian P, Dehghani A, El-Kattan Y, Lin TH, Hutchison TL, Elliott AJ, Parker CD et al (2000) BCX-1812 (RWJ-270201): Discovery of a novel, highly potent, orally active, and selective influenza neuraminidase inhibitor through structure-based drug design. J Med Chem 43:3482–3486
35. Streicher H, Stanley M (2011) The development of non-carbohydrate-based influenza virus sialidase inhibitors. In: von Itzstein M (ed) Influenza virus sialidase: a drug discovery target. Springer, Heidelberg
36. Meindl P, Tuppy H (1969) Verfahren zur Herstellung neuer 2-Desoxy-2,3-dehydro-*N*-acylneuraminsäuren. DE 1493249. (CAN 78:30153)
37. Haverkamp J, Schauer R, Wember M, Farriaux J-P, Kamerling JP, Versluis C, Vliegenthart FG (1976) Neuraminic acid derivatives newly discovered in humans. Hoppe-Seyler's Z Physiol Chem 357:1699–1705
38. Holzer CT, von Itzstein M, Jin B, Pegg MS, Stewart WP, Wu WY (1993) Inhibition of sialidases from viral, bacterial and mammalian sources by analogues of 2-deoxy-2,3-didehydro-*N*-acetylneuraminic acid modified at the C-4 position. Glycoconjugate J 10:40–44
39. Miller CA, Wang P, Flashner M (1978) Mechanism of *Arthrobacter sialophilus* neuraminidase: the binding of substrates and transition-state analogs. Biochem Biophys Res Commun 83:1479–1487
40. Furuhata K, Sato S, Goto M, Takayanagi H, Ogura H (1988) Studies on sialic acids. VII. The crystal and molecular structure of *N*-acetyl-2,3-dehydro-2-deoxyneuraminic acid. Chem Pharm Bull 36:1872–1876
41. Zbiral E, Brandstetter HH, Christian R, Schauer R (1987) Structural variations of *N*-acetylneuraminic acid. 7. Synthesis of the C-7-, C-8- and C-7, -8-side chain epimers of 2-deoxy-2,3-didehydro-*N*-acetylneuraminic acid and their behavior towards sialidase from *Vibrio cholerae*. Liebigs Ann Chem: 781–786
42. Zbiral E, Schreiner E, Christian R, Kleineidam RG, Schauer R (1989) Structural variations of *N*-acetylneuraminic acid. 10. Synthesis of 2,7-, 2,8-, and 2,9-dideoxy- and 2,4,7-trideoxy-2,

3-didehydro-*N*-acetylneuraminic acids and their behavior towards sialidase from *Vibrio cholerae*. Liebigs Ann Chem: 159–165
43. Schreiner E, Zbiral E, Kleineidam RG, Schauer R (1991) Structural variations on *N*-acetylneuraminic acid, 20. Synthesis of some 2,3-didehydro-2-deoxysialic acids structurally varied at C-4 and their behavior towards sialidase from *Vibrio cholerae*. Liebigs Ann Chem: 129–134
44. Schreiner E, Zbiral E, Kleineidam RG, Schauer R (1991) Structural variations on *N*-acetylneuraminic acid. Part 21. 2,3-Didehydro-2-deoxysialic acids structurally varied at C-5 and their behavior toward the sialidase from *Vibrio cholerae*. Carbohydr Res 216:61–66
45. Vasella A, Wyler R (1991) Synthesis of a phosphonic acid analog of *N*-acetyl-2,3-didehydro-2-deoxyneuraminic acid, an inhibitor of *Vibrio cholerae* sialidase. Helv Chim Acta 74:451–463
46. Zbiral E (1992) Synthesis of sialic acid analogues and their behaviour towards the enzymes of sialic acid metabolism and hemagglutinin X-31 of influenza A-virus. In: Ogura H, Hasegawa A, Suami T (eds) Carbohydrates. Synthetic methods and applications in medicinal chemistry. VCH, Weinheim, pp 304–339
47. Palese P, Schulman JL, Bodo G, Meindl P (1974) Inhibition of influenza and parainfluenza virus replication in tissue culture by 2-deoxy-2,3-dehydro-*N*-trifluoroacetylneuraminic acid (FANA). Virology 59:490–498
48. Palese P, Compans RW (1976) Inhibition of influenza virus replication in tissue culture by 2-deoxy-2,3-dehydro-*N*-trifluoroacetylneuraminic acid (FANA): mechanism of action. J Gen Virol 33:159–163
49. Palese P, Schulman JL (1977) Inhibitors of viral neuraminidase as potential antiviral drugs. In: Oxford JS (ed) Chemoprophylaxis and viral infection of the upper respiratory tract. CRC Press, Cleveland, Ohio, pp 189–202
50. Nöhle U, Beau JM, Schauer R (1982) Uptake, metabolism and excretion of orally and intravenously administered, double-labeled *N*-glycoloylneuraminic acid and single-labeled 2-deoxy-2,3-dehydro-*N*-acetylneuraminic acid in mouse and rat. Eur J Biochem 126:543–548
51. Varghese JN, Laver WG, Colman PM (1983) Structure of the influenza virus glycoprotein antigen neuraminidase at 2.9Å resolution. Nature 303:35–40
52. Varghese JN, Colman PM (1991) Three-dimensional structure of the neuraminidase of influenza virus A/Tokyo/3/67 at 2.2Å resolution. J Mol Biol 221:473–486
53. Goodford PJ (1985) A computational procedure for determining energetically favorable binding sites on biologically important macromolecules. J Med Chem 28:849–857
54. Driguez PA, Barrere B, Quash G, Doutheau A (1994) Synthesis of transition-state analogs as potential inhibitors of sialidase from influenza-virus. Carbohydr Res 262:297–310
55. von Itzstein M, Wu W-Y, Jin B (1994) The synthesis of 2,3-didehydro-2,4-dideoxy-4-guanidinyl-*N*-acetylneuraminic acid: a potent influenza virus sialidase inhibitor. Carbohydr Res 259:301–305
56. Hata K, Koseki K, Yamaguchi K, Moriya S, Suzuki Y, Yingsakmongkon S, Hirai G, Sodeoka M, von Itzstein M, Miyagi T (2008) Limited inhibitory effects of oseltamivir and zanamivir on human sialidases. Antimicrob Agents Chemother 52:3484–3491
57. Chavas LMG, Kato R, Suzuki N, von Itzstein M, Mann MC, Thomson RJ, Dyason JC, McKimm-Breschkin J, Fusi P, Tringali C et al (2010) Complexity in influenza virus targeted drug design: interaction with human sialidases. J Med Chem 53:2998–3002
58. Hart GJ, Bethell RC (1995) 2,3-Didehydro-2,4-dideoxy-4-guanidino-*N*-acetyl-D-neuraminic acid (4-guanidino-Neu5Ac2en) is a slow-binding inhibitor of sialidase from both influenza A virus and influenza B virus. Biochem Mol Biol Int 36:695–703
59. Bantia S, Parker CD, Ananth SL, Horn LL, Andries K, Chand P, Kotian PL, Dehghani A, El Kattan Y, Lin T et al (2001) Comparison of the anti-influenza virus activity of RWJ-270201 with those of oseltamivir and zanamivir. Antimicrob Agents Chemother 45: 1162–1167

60. Kati WM, Montgomery D, Carrick R, Gubareva L, Maring C, McDaniel K, Steffy K, Molla A, Hayden F, Kempf D et al (2002) In vitro characterization of A-315675, a highly potent inhibitor of A and B strain influenza virus neuraminidases and influenza virus replication. Antimicrob Agents Chemother 46:1014–1021
61. Govorkova EA, Leneva IA, Goloubeva OG, Bush K, Webster RG (2001) Comparison of efficacies of RWJ-270201, zanamivir, and oseltamivir against H5N1, H9N2, and other avian influenza viruses. Antimicrob Agents Chemother 45:2723–2732
62. Waghorn SL, Goa KL (1998) Zanamivir. Drugs 55:721–725
63. Elliott M (2001) Zanamivir: from drug design to the clinic. Phil Trans R Soc Lond B Biol Sci 356:1885–1893
64. von Itzstein M (2007) The war against influenza: discovery and development of sialidase inhibitors. Nat Rev Drug Discov 6:967–974
65. von Itzstein M, Thomson R (2009) Anti-influenza drugs: the development of sialidase inhibitors. Handb Exp Pharmacol 189:111–154
66. Woods JM, Bethell RC, Coates JAV, Healy N, Hiscox SA, Pearson BA, Ryan DM, Ticehurst J, Tilling J, Walcott SM et al (1993) 4-Guanidino-2,4-dideoxy-2,3-dehydro-*N*-acetylneuraminic acid is a highly effective inhibitor both of the sialidase (neuraminidase) and of growth of a wide range of influenza-A and influenza-B viruses *in vitro*. Antimicrob Agents Chemother 37:1473–1479
67. Varghese JN, Epa VC, Colman PM (1995) Three-dimensional structure of the complex of 4-guanidino-Neu5Ac2en and influenza virus neuraminidase. Protein Sci 4:1081–1087
68. Morris SJ, Price GE, Barnett JM, Hiscox SA, Smith H, Sweet C (1999) Role of neuraminidase in influenza virus-induced apoptosis. J Gen Virol 80:137–146
69. Monti E, Preti A, Venerando B, Borsani G (2002) Recent development in mammalian sialidase molecular biology. Neurochem Res 27:649–663
70. Calfee DP, Peng AW, Cass LM, Lobo M, Hayden FG (1999) Safety and efficacy of intravenous zanamivir in preventing experimental human influenza A virus infection. Antimicrob Agents Chemother 43:1616–1620
71. Kidd IM, Down J, Nastouli E, Shulman R, Grant PR, Howell DC, Singer M (2009) H1N1 pneumonitis treated with intravenous zanamivir. Lancet 374:1036
72. Gaur AH, Bagga B, Barman S, Hayden R, Lamptey A, Hoffman JM, Bhojwani D, Flynn PM, Tuomanen E, Webby R (2010) Intravenous zanamivir for oseltamivir-resistant 2009 H1N1 influenza. N Engl J Med 362:88–89
73. Buchy P (2011) Clinical experience with influenza virus sialidase inhibitors. In: von Itzstein M (ed) Influenza virus sialidase: a drug discovery target. Springer, Heidelberg
74. Ryan DM, Ticehurst J, Dempsey M, Penn CR (1994) Inhibition of influenza virus replication in mice by GG167 (4-guanidino-2,4-dideoxy-2,3-dehydro-*N*-acetylneuraminic acid) is consistent with extracellular actiivity of viral neuraminidase (sialidase). Antimicrob Agents Chemother 10:2270–2275
75. Liu Z-Y, Wang B, Zhao L-X, Li Y-H, Shao H-Y, Yi H, You X-F, Li Z-R (2007) Synthesis and anti-influenza activities of carboxyl alkoxyalkyl esters of 4-guanidino-Neu5Ac2en (zanamivir). Bioorg Med Chem Lett 17:4851–4854
76. Honda T, Masuda T, Yoshida S, Arai M, Kobayashi Y, Yamashita M (2002) Synthesis and anti-influenza virus activity of 4-guanidino-7-substituted Neu5Ac2en derivatives. Bioorg Med Chem Lett 12:1921–1924
77. Yamashita M (2004) R-118958, a unique anti-influenza agent showing high efficacy for both prophylaxis and treatment after a single administration: from the in vitro stage to phase I study. Int Congress Ser 1263:38–42
78. Yamashita M, Tomozawa T, Kakuta M, Tokumitsu A, Nasu H, Kubo S (2009) CS-8958, a prodrug of the new neuraminidase inhibitor R-125489, shows long-acting anti-influenza virus activity. Antimicrob Agents Chemother 53:186–192

79. Honda T, Kubo S, Masuda T, Arai M, Kobayashi Y, Yamashita M (2009) Synthesis and in vivo influenza virus-inhibitory effect of ester prodrug of 4-guanidino-7-*O*-methyl-Neu5Ac2en. Bioorg Med Chem Lett 19:2938–2940
80. Koyama K, Takahashi M, Oitate M, Nakai N, Takakusa H, S-i M, Okazaki O (2009) CS-8958, a prodrug of the novel neuraminidase inhibitor R-125489, demonstrates a favorable long-retention profile in the mouse respiratory tract. Antimicrob Agents Chemother 53:4845–4851
81. Kubo S, Tomozawa T, Kakuta M, Tokumitsu A, Yamashita M (2010) Laninamivir prodrug CS-8958, a long-acting neuraminidase inhibitor, shows superior anti-influenza virus activity after a single administration. Antimicrob Agents Chemother 54:1256–1264
82. Kiso M, Kubo S, Ozawa M, Le QM, Nidom CA, Yamashita M, Kawaoka Y (2010) Efficacy of the new neuraminidase inhibitor CS-8958 against H5N1 influenza viruses. PLoS Pathog 6: e1000786
83. Honda T, Masuda T, Yoshida S, Arai M, Kaneko S, Yamashita M (2002) Synthesis and anti-Influenza virus activity of 7-*O*-alkylated derivatives related to zanamivir. Bioorg Med Chem Lett 12:1925–1928
84. Andrews DM, Cherry PC, Humber DC, Jones PS, Keeling SP, Martin PF, Shaw CD, Swanson S (1999) Synthesis and influenza virus sialidase inhibitory activity of analogues of 4-guanidino-Neu5Ac2en (Zanamivir) modified in the glycerol side-chain. Eur J Med Chem 34:563–574
85. Watson KG, Cameron R, Fenton RJ, Gower D, Hamilton S, Jin B, Krippner GY, Luttick A, McConnell D, MacDonald SJ et al (2004) Highly potent and long-acting trimeric and tetrameric inhibitors of influenza virus neuraminidase. Bioorg Med Chem Lett 14:1589–1592
86. Honda T, Yoshida S, Arai M, Masuda T, Yamashita M (2002) Synthesis and anti-influenza evaluation of polyvalent sialidase inhibitors bearing 4-guanidino-Neu5Ac2en derivatives. Bioorg Med Chem Lett 12:1929–1932
87. Macdonald SJ, Watson KG, Cameron R, Chalmers DK, Demaine DA, Fenton RJ, Gower D, Hamblin JN, Hamilton S, Hart GJ et al (2004) Potent and long-acting dimeric inhibitors of influenza virus neuraminidase are effective at a once-weekly dosing regimen. Antimicrob Agents Chemother 48:4542–4549
88. Macdonald SJ, Cameron R, Demaine DA, Fenton RJ, Foster G, Gower D, Hamblin JN, Hamilton S, Hart GJ, Hill AP et al (2005) Dimeric zanamivir conjugates with various linking groups are potent, long-lasting inhibitors of influenza neuraminidase including H5N1 avian influenza. J Med Chem 48:2964–2971
89. Wen W-H, Lin M, Su C-Y, Wang S-Y, Cheng Y-SE, Fang J-M, Wong C-H (2009) Synergistic effect of zanamivir–porphyrin conjugates on inhibition of neuraminidase and inactivation of influenza virus. J Med Chem 52:4903–4910
90. Sun X-L (2007) Recent anti-influenza strategies in multivalent sialyloligosaccharides and sialylmimetics approaches. Curr Med Chem 14:2304–2313
91. Haldar J, Alvarez de Cienfuegos L, Tumpey TM, Gubareva LV, Chen J, Klibanov AM (2010) Bifunctional polymeric inhibitors of human influenza A viruses. Pharm Res 27:259–263
92. Bamford MJ (1995) Neuraminidase inhibitors as potential anti-influenza drugs. J Enzyme Inhib 10:1–16
93. Thomson RJ, von Itzstein M (2003) *N*-Acetylneuraminic acid derivatives and mimetics as anti-influenza agents. In: Wong C-H (ed) Carbohydrate-based drug discovery. Wiley-VCH, Weinheim, pp 831–862
94. Streicher H (2004) Synthesis and evaluation as sialidase inhibitors of xylo-configured cyclohexene-phosphonates carrying glycerol side-chain mimics. Bioorg Med Chem Lett 14:361–364
95. Shie J-J, Fang J-M, Wang S-Y, Tsai K-C, Cheng Y-SE, Yang A-S, Hsiao S-C, Su C-Y, Wong C-H (2007) Synthesis of Tamiflu and its phosphonate congeners possessing potent anti-influenza activity. J Am Chem Soc 129:11892–11893

96. Carbain B, Collins PJ, Callum L, Martin SR, Hay AJ, McCauley J, Streicher H (2009) Efficient synthesis of highly active phospha-isosteres of the influenza neuraminidase inhibitor oseltamivir. ChemMedChem 4:335–337
97. Chandler M, Bamford MJ, Conroy R, Lamont B, Patel B, Patel VK, Steeples IP, Storer R, Weir NG, Wright M et al (1995) Synthesis of the potent influenza neuraminidase inhibitor 4-guanidino-Neu5Ac2en. X-Ray molecular structure of 5-acetamido-4-amino-2,6-anhydro-3,4,5-trideoxy-D-*erythro*-L-*gluco*-nononic acid. J Chem Soc Perkin Trans 1:1173–1180
98. Ikeda K, Sano K, Ito M, Saito M, Hidari K, Suzuki T, Suzuki Y, Tanaka K (2001) Synthesis of 2-deoxy-2,3-didehydro-*N*-acetylneuraminic acid analogues modified at the C-4 and C-9 positions and their behavior towards sialidase from influenza virus and pig liver membrane. Carbohydr Res 330:31–41
99. Wen W-H, Wang S-Y, Tsai K-C, Cheng Y-SE, Yang A-S, Fang J-M, Wong C-H (2010) Analogs of zanamivir with modified C4-substituents as the inhibitors against the group-1 neuraminidases of influenza viruses. Bioorg Med Chem 18:4074–4084
100. Li J, Zheng M, Tang W, He P-L, Zhu W, Li T, Zuo J-P, Liu H, Jiang H (2006) Syntheses of triazole-modified zanamivir analogs via click chemistry and anti-AIV activities. Bioorg Med Chem Lett 16:5009–5013
101. Bamford MJ, Pichel JC, Husman W, Patel B, Storer R, Wier NG (1995) Synthesis of 6-, 7- and 8-carbon sugar analogs of potent anti-influenza 2,3-didehydro-2,3-dideoxy-*N*-acetylneuraminic acid derivatives. J Chem Soc Perkin Trans 1:1181–1187
102. Masuda T, Shibuya S, Arai M, Yoshida S, Tomozawa T, Ohno A, Yamashita M, Honda T (2003) Synthesis and anti-influenza evaluation of orally active bicyclic ether derivatives related to zanamivir. Bioorg Med Chem Lett 13:669–673
103. Smith PW, Sollis SL, Howes PD, Cherry PC, Cobley KN, Taylor H, Whittington AR, Scicinski J, Bethell RC, Taylor N et al (1996) Novel inhibitors of influenza sialidases related to GG167. Structure-actvity, crystallographic and molecular dynamics studies with 4-*H*-pyran-2-carboxylic acid 6-carboxamides. Bioorg Med Chem Lett 6:2931–2936
104. Smith PW, Sollis SL, Howes PD, Cherry PC, Starkey ID, Cobley KN, Weston H, Scicinski J, Merritt A, Whittington A et al (1998) Dihydropyrancarboxamides related to zanamivir: a new series of inhibitors of influenza virus sialidases. 1. Discovery, synthesis, biological activity, and structure-activity relationships of 4-guanidino- and 4-amino-4 *H*-pyran-6-carboxamides. J Med Chem 41:787–797
105. Wyatt PG, Coomber BA, Evans DN, Jack TI, Fulton HE, Wonacott AJ, Colman P, Varghese J (2001) Sialidase inhibitors related to zanamivir. Further SAR studies of 4-amino-4 *H*-pyran-2-carboxylic acid-6-propylamides. Bioorg Med Chem Lett 11:669–673
106. Taylor NR, Cleasby A, Singh O, Skarzynski T, Wonacott AJ, Smith PW, Sollis SL, Howes PD, Cherry PC, Bethell R et al (1998) Dihydropyrancarboxamides related to zanamivir: a new series of inhibitors of influenza virus sialidases. 2. Crystallographic and molecular modeling study of complexes of 4-amino-4 *H*-pyran-6-carboxamides and sialidase from influenza virus types A and B. J Med Chem 41:798–807
107. Vorwerk S, Vasella A (1998) Carbocyclic analogs of *N*-acetyl-2,3-didehydro-2-deoxy D-neuraminic acid (Neu5Ac2en, DANA): synthesis and inhibition of viral and bacterial neuraminidases. Angew Chem Int Ed Engl 37:1732–1734
108. Lew W, Chen X, Kim CU (2000) Discovery and development of GS 4104 (oseltamivir): an orally active influenza neuraminidase inhibitor. Curr Med Chem 7:663–672
109. Kim CU, Lew W, Williams MA, Liu H, Zhang L, Swaminathan S, Bischofberger N, Chen MS, Mendel DB, Tai CY et al (1997) Influenza neuraminidase inhibitors possessing a novel hydrophobic interaction in the enzyme active site: design, synthesis, and structural analysis of carbocyclic sialic acid analogues with potent anti-influenza activity. J Am Chem Soc 119:681–690
110. Smith PW, Robinson JE, Evans DN, Sollis SL, Howes PD, Trivedi N, Bethell RC (1999) Sialidase inhibitors related to zanamivir: synthesis and biological evaluation of 4 *H*-pyran 6-ether and ketone. Bioorg Med Chem Lett 9:601–604

111. Florio P, Thomson RJ, Alafaci A, Abo S, von Itzstein M (1999) Synthesis of Δ-4-β-D-glucopyranosiduronic acids as mimetics of 2,3-unsaturated sialic acids for sialidase inhibition. Bioorg Med Chem Lett 9:2065–2068
112. Mann MC, Islam T, Dyason JC, Florio P, Trower CJ, Thomson RJ, von Itzstein M (2006) Unsaturated *N*-acetyl-D-glucosaminuronic acid glycosides as inhibitors of influenza virus sialidase. Glycoconjugate J 23:127–133
113. Mann MC, Thomson RJ, Dyason JC, McAtamney S, von Itzstein M (2006) Modelling, synthesis and biological evaluation of novel glucuronide-based probes of *Vibrio cholerae* sialidase. Bioorg Med Chem 14:1518–1537
114. Kok GB, Campbell M, Mickey B, von Itzstein M (1996) Synthesis and biological evaluation of sulfur isosteres of the potent influenza virus sialidase inhibitors 4-amino-4-deoxy- and 4-deoxy-4-guanidino-Neu5Ac2en. J Chem Soc Perkin Trans 1:2811–2815
115. Kerrigan SA, Pritchard RG, Smith PW, Stoodley RJ (2001) Synthesis of (4R*,5 S*)-5-(acetylamino)-4-(diethylcarbamoyl)-5,6-dihydro-4 H-pyran-2-carboxylic acid and its inhibitory action against influenza virus sialidases. Tetrahedron Lett 42:8889–8892
116. Meanwell NA, Krystal M (1996) Taking aim at a moving target - Inhibitors of influenza virus.1. Virus adsorption, entry and uncoating. Drug Discov Today 1:316–324
117. Meanwell NA, Krystal M (1996) Taking aim at a moving target - Inhibitors of influenza virus. 2. Viral replication, packaging and release. Drug Discov Today 1:388–397
118. Lagoja IM, De Clercq E (2008) Anti-influenza agents: synthesis and mode of action. Med Res Rev 28:1–38
119. Hayden F (2009) Developing new antiviral agents for influenza treatment: what does the future hold? Clin Infect Dis 48:S3–S13
120. Furuta Y, Takahashi K, Shiraki K, Sakamoto K, Smee DF, Barnard DL, Gowen BB, Julander JG, Morrey JD (2009) T-705 (favipiravir) and related compounds: Novel broad-spectrum inhibitors of RNA viral infections. Antiviral Res 82:95–102
121. Triana-Baltzer GB, Babizki M, Chan MCW, Wong ACN, Aschenbrenner LM, Campbell ER, Li Q-X, Chan RWY, Peiris JS, Nicholls JM et al (2010) DAS181, a sialidase fusion protein, protects human airway epithelium against influenza virus infection: an *in vitro* pharmacodynamic analysis. J Antimicrob Chemother 65:275–284
122. Triana-Baltzer GB, Gubareva LV, Nicholls JM, Pearce MB, Mishin VP, Belser JA, Chen L-M, Chan RWY, Chan MCW, Hedlund M et al (2009) Novel pandemic influenza A(H1N1) viruses are potently inhibited by DAS181, a sialidase fusion protein. PLoS One 4:e7788

The Development of Non-Carbohydrate-Based Influenza Virus Sialidase Inhibitors

Hansjörg Streicher and Mathew Stanley

Abstract The chapter summarises the development of influenza virus sialidase inhibitors that are not structurally derived from carbohydrates, for instance sialic acid. The classification of inhibitor structures, such as the approved drugs oseltamivir and peramivir, is made based on the type of scaffold used, from six-membered aromatic systems to five-membered carbocylic scaffolds. For the most important examples, milestones and unexpected results during their development as well as the respective syntheses are discussed. The last two sections describe natural products with influenza sialidase inhibitory activity which might serve as lead structures in the future and, in brief, discuss the increasing number of *in silico* studies in the field.

1 Introduction

Two of the three influenza virus sialidase (NA) inhibitors, oseltamivir carboxylate **1** and peramivir **2**, which are currently approved for treatment of infection by the influenza virus, are non-carbohydrate-based compounds (in case of **1** as its ethyl ester prodrug) [1, 2]. The third, zanamivir **3** (also preceeding chapters), is derived from sialic acid and the development of all three inhibitors can be traced back to the long-known generic sialidase inhibitor and lead compound DANA **4** (2-deoxy-2.3-didehydro-*N*-acetylneuraminic acid, Neu5Ac2en) (Fig. 1) [3].

DANA **4** is the product of the reaction of a sialyl oxocarbenium ion intermediate generated during sialoside cleavage by sialidase eliminating a proton instead of being trapped by water as acceptor. It inhibits influenza virus sialidase in the micromolar range, i.e. it is much more active than the standard non-hydrolysable sialosides such as C- or S-sialosides. This potency is attributed to it mimicking

H. Streicher (✉) • M. Stanley
Department of Chemistry and Biochemistry, University of Sussex, Brighton BN1 9QG UK
e-mail: h.streicher@sussex.ac.uk

M. von Itzstein (ed.), *Influenza Virus Sialidase - A Drug Discovery Target*,
Milestones in Drug Therapy, DOI 10.1007/978-3-7643-8927-7_6,

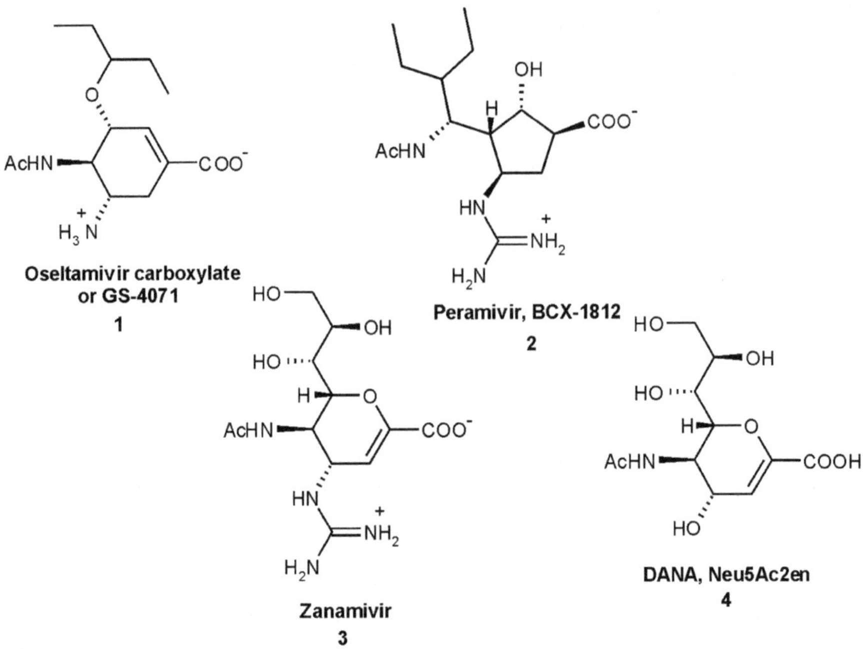

Fig. 1 Structures of lead compound 2-deoxy-2.3-didehydro-*N*-acetylneuraminic acid (DANA, Neu5Ac2en) and commercialised inhibitors zanamivir, oseltamivir and peramivir

a flattened half-chair transition state in sialoside hydrolysis [4]. Consequently, simulating its more flexible conformation compared to sialosides has become a dominant theme in the search for improved inhibitors. Zanamivir **3** is essentially a DANA molecule bearing a guanidino group in position 4 which adds a strong additional ionic interaction with the sialidase active site [5]. Although zanamivir **3**, which is classified as carbohydrate-(sialic acid) based inhibitor, entered the market first and is as potent as the other two drugs, the market was dominated by oseltamivir carboxylate **1** during the recent swine flu pandemic. This is due to improved pharmacokinetic properties that have been achieved through intense, decade-long drug-development programs, efforts which are still ongoing.

In this chapter, we attempt to give an overview of the development of non-carbohydrate-based influenza virus sialidase inhibitors during the past 40 years. We start with a section concerned with inhibitors that have been designed rationally to mimic sialic acid, the sialosyl cation or indeed DANA but with improved properties [6–11]. We have organised this section on the basis of the type of scaffold which was used to replace the pyran/pyranose system and we hope to show that compromises have to be made with respect to potency versus cost of synthesis. The second section will include influenza sialidase inhibitors that are structurally unrelated to sialic acid and which have in many cases been isolated from natural sources. Their modes of inhibition are often different from those of designed sialyl mimetics but nevertheless they can serve as valuable leads for new generations of drugs. With the third short and not comprehensive section, we intend to summarise the relatively new and growing field of computer-aided design of virtual inhibitors and inhibitor libraries. These are typically structure based but often do not include syntheses or assay-derived inhibition data.

2 Non-carbohydrate-Based Influenza Virus Sialidase Inhibitors Structurally Related to Sialic Acids and Designed to Mimic Sialic Acids

The carboxylate group can probably be considered the most important motif for recognition and binding of substrates and inhibitors by influenza virus sialidase, indeed by all sialidases. This is also reflected by the fact that in the large family of sialic acids, modifications have been found at every position except for the carboxylate [12]. Therefore, preserving or simulating this acidic group, which is negatively charged under physiological conditions and complemented by the arginine triad in the enzyme's active site (see also preceeding chapters), is an essential feature running through most of the following approaches. In order to aid understanding the binding of the various structures discussed later, Fig. 2 summarises some important interactions of oseltamivir carboxylate **1** bound to the influenza virus A sialidase (NA(A)) active site [13] and lists the subsites important for discussion of binding modes.

2.1 Influenza Virus Sialidase Inhibitors Based on Aromatic and Heteroaromatic Scaffolds

Medicinal chemistry attempts to improve the pharmacokinetic and pharmacodynamic properties of a lead compound as well as reduce costs through simplification of the structure and thus the synthesis. From this standpoint, the simplest mimetics of Neu5Ac or DANA are aromatic and heteroaromatic systems such as benzoic acids, pyridinecarboxylic acids and the like [6, 9, 10]. Mimetics of the other important structural features of Neu5Ac can then be introduced by standard aromatic/heterocyclic chemistry, often involving a minimum of or no stereochemistry.

Figure 3 summarises some of the scaffold types used and substitution patterns generated during the synthesis of libraries of inhibitors in both academic and industrial environments, mainly in the 1990s and early 2000s. The aromatic analogue of zanamivir **5** (R = CH(OH)CH(OH)CH_2OH) was not effective [14] but inhibition in the low micromolar range was achieved with the reduced structures BCX-140 **6** (BANA, R = H, 2.5 μM for NA(A)) [15–17] or pyridinecarboxylic acid **7** (IC_{50} = 4 μM for NA(A) (H1N9))[18]. The guanidino group in **6**, however, replaces the glycerol side chain in S5 rather than undergoing the zanamivir-type ionic interactions with S2. Efforts to reverse this interaction by varying this

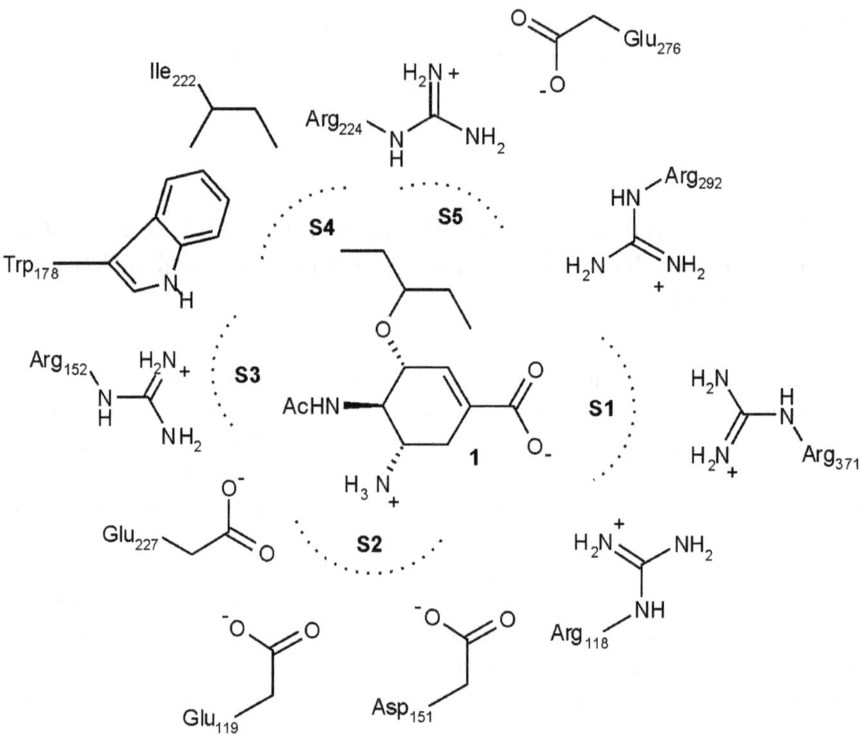

Fig. 2 Oseltamivir carboxylate **1** bound to the influenza virus A sialidase active site

substituent resulted at best in not changing the affinity as did studies in which 2-aminoimidazoles replaced the guanidine [19, 20]. The only significant improvement in this mode of binding came when hydroxypyrrolidinones **8** and **9** which replaced the acetamide in S3 but also interacted with S2, inhibited as effectively as BCX-140 with essentially the same orientation in the active site [21–24]. This allowed the occupation of S5 with an additional hydrophobic substituent (in **9**) and improved inhibition of NA(A) (but not B) to an IC_{50} of 0.048 μM [24]. The glycerol binding pocket undergoes a rearrangement allowing for more hydrophobic interactions (see later for oseltamivir). More recently, a study reporting inhibition of NA(A) (H3N2) in the 40–50 nM range was published [25]. The authors used *p*-aminosalicylic acid as a scaffold for molecules of type **10**. This design was supported by QSAR studies. An IC_{50} value of 0.038 μM was reported for compound **11** and compared to oseltamivir's value of 0.021 μM in the same assay format. Interestingly, in this compound the carboxylate is methylated and thus not available for the crucial ionic interaction with the arginine triad. It will be interesting to see whether this inhibition is competitive and to what extent the mode of binding of **11** differs from, for example, the benzoic acid **9** (Fig. 3). Although the methyl ester of sialic acid can inhibit sialidase, it does so much more weakly than the free acid [26]. The assay conditions need to be carefully examined to ensure that **11** is not acting as

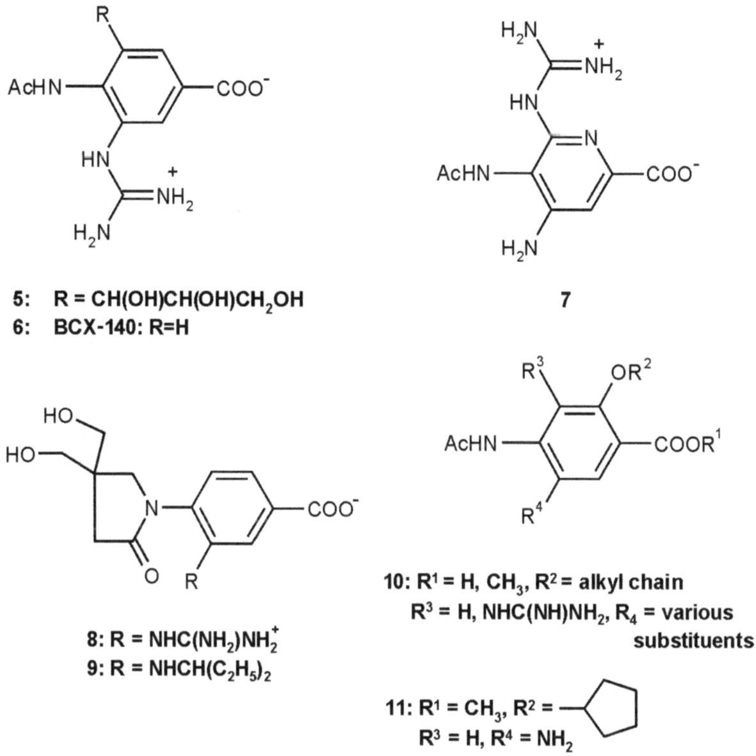

Fig. 3 Structures of influenza sialidase inhibitors based on aromatic scaffolds

pro-inhibitor which is hydrolysed to the active compound under the basic conditions used for fluorimetric analysis [25].

2.2 *Oseltamivir and Inhibitors Based on (Non-aromatic) Six-Membered Carbocyclic Scaffolds*

2.2.1 From DANA to Oseltamivir

The development of oseltamivir was inspired by the fact that one problem of sialic acid-based inhibitors like zanamivir **3** or DANA **4** is their rather polar nature. Apart from the basic group in position 4, the acetamide in position 5 and the essential carboxylate, this polarity is mainly due to the ring oxygen and the polar glycerol side chain. The replacement of both the ring oxygen and polar glycerol side chain was supported by a series of key findings. Firstly, it was realised that the glycerol side chain could be replaced by substantially more hydrophobic moieties (for instance, dialkyl carboxamide groups in zanamivir-type compounds like **12** or the

3-pentylamine in the aromatic series) [24, 27–30]. This is made possible by a shift of a glutamic acid (Glu 276) side chain to form an ionic bond with Arg-224 and thus rendering the glycerol-binding subsite more hydrophobic. All subsequent oseltamivir-type inhibitors take advantage of this effect. Secondly, once the ring oxygen was considered not to be essential [31], investigation of the methylene isosteres **13** and **14** of DANA indicated that **14** in which the double bond position does not correspond to that in DANA is strongly preferred (Fig. 4) [32–34].

Thirdly, introduction of a hydroxyl group instead of C-7 in **15** reduced (but did not abolish) activity. In turn this facilitated the systematic optimisation of the side chain mimic in **16** with the 3-pentyl ether of **1** being the final result (Fig. 4) [34, 35]. In contrast to zanamivir **3**, the amino group proved equally potent to the guanidine in binding to S2.

Further miscellaneous modifications of the substituents on the cyclohexene scaffold were investigated [36], but proved inferior either in NA inhibition or plaque reduction assays. Examples are alkylamine-[37] or cycloalkylamine [38] side chain mimics, thia and carba isosteres of alkoxy substituents [39], other acyl or sulfonamide replacements of the acetamide [36] or substituents at the double bond [40], i.e. corresponding to the ring oxygen position in DANA.

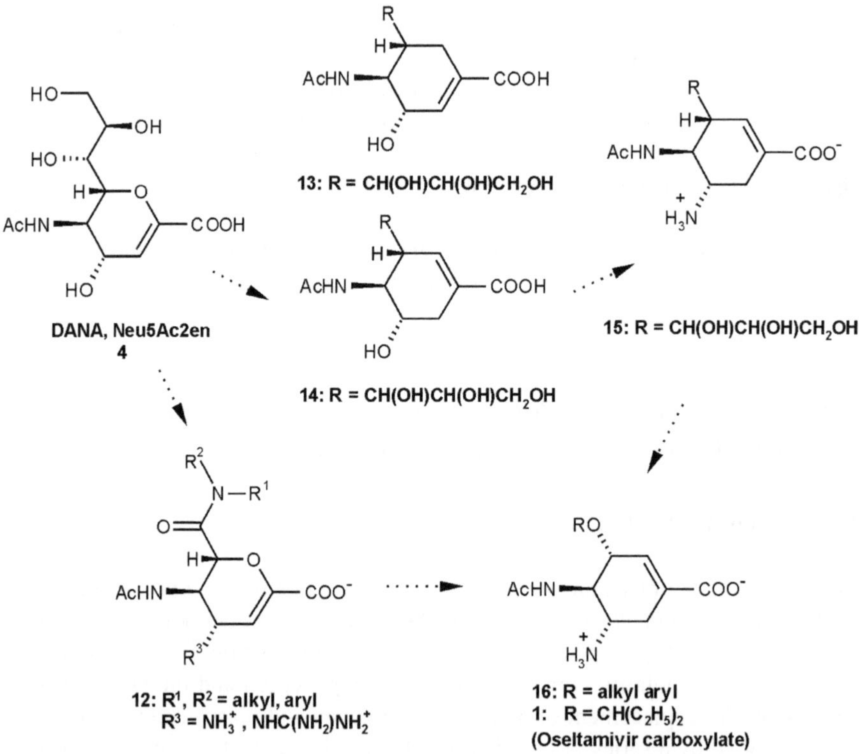

Fig. 4 Important findings that influenced the development of oseltamivir carboxylate **1** from DANA **4**

Oseltamivir carboxylate **1** itself did not show a marked improvement in bioavailability compared to zanamivir, which has to be administered as a nasal spray. The ethyl ester prodrug **1a**, which is suitable for oral administration and is hydrolysed by endogenous esterases, can be used and this was approved by the FDA in 1999 (Scheme 1). It has been marketed since, in a phosphate salt formulation, by F. Hoffmann-LaRoche Ltd. under the trade name Tamiflu™. Sales in 2009, which included a spike due to the swine flu crisis, were well in excess of $2 billion.

2.2.2 Oseltamivir Syntheses

The initial syntheses of oseltamivir by Gilead Sciences were based on the chiral pool and used (−)-quinic acid **17** and (−)-shikimic **20** acid as the starting materials (Scheme 2) [34, 41].

The industrial synthesis which is currently used by Roche has remained remarkably close to this even employing an unpopular, large-scale, azide-based synthetic step that required outsourcing. Improvements, modifications and alternatives to this synthesis have been extensively reviewed by the team at Roche since (Scheme 3) [42, 43].

Very recently, the Roche group have published a shorter, eight-step, improvement of their oseltamivir synthesis from shikimic acid. This has simplified workup procedures and does not need protecting group modification or chromatographic separations. The overall yield was 20% from shikimic acid [44].

For several reasons, intense competition for the most effective total synthesis of oseltamivir has developed over recent years: The limited availability of the starting material shikimic acid, which is either isolated from star anise or produced in a fermentation process by a recombinant *E. coli* strain. The emergence of the more recent avian flu (H5N1) and swine flu (H1N1) threat with Tamiflu™ being stockpiled as the first line of defence led to an increase in revenue. Last but not least, the cost reduction that can be expected from a more efficient industrial process of a successful drug is significant. Oseltamivir is a relatively small

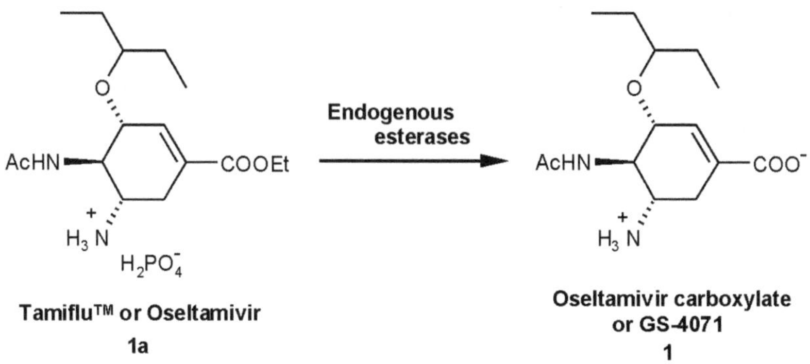

Scheme 1 The oseltamivir prodrug concept

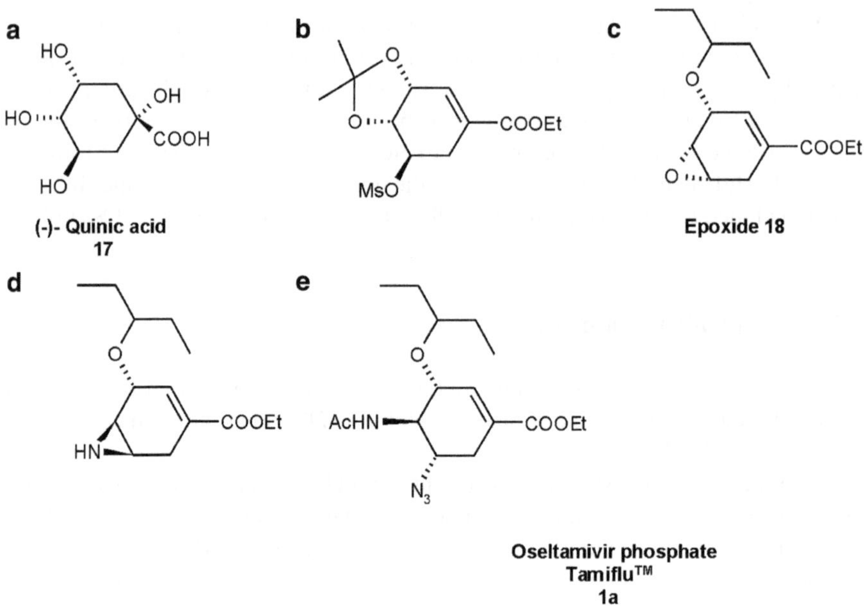

Scheme 2 The scaled-up synthesis of oseltamivir **1a** by Gilead Sciences. (**a**) 2,2-DMP, TsOH; EtOH, EtONa; MsCl, NEt_3; SO_2Cl_2, pyridine; 30% overall. (**b**) 3-pentanone, $HClO_4$, BH_3-Me_2S, TMSOTf; $KHCO_3$, aq. EtOH (60–72%). (**c**) NaN_3, NH_4Cl, aq. EtOH; PMe_3 (**d**) NaN_3, NH_4Cl, DMF; Ac_2O. (**e**) Ra-Ni, H_2, EtOH; 85% H_3PO_4

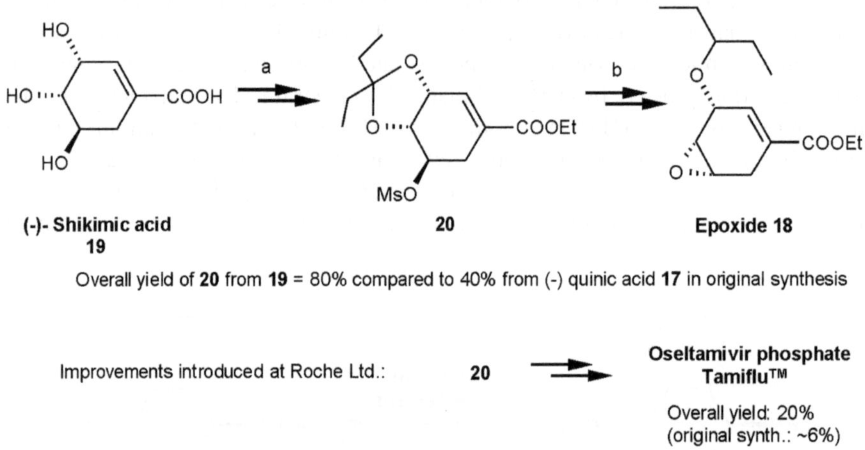

Scheme 3 Selected improvements made during the development of the industrial synthesis of oseltamivir phosphate. (**a**) EtOH, $SOCl_2$; 3-pentanone, CF_3SO_3H; MsCl, NEt_3. (**b**) Et_3SiH, $TiCl_4$; $NaHCO_3$, H_2O, EtOH

molecule, but densely functionalised, and has meanwhile proven to be a formidable target for many leading laboratories specialising in total synthesis. This matter has been comprehensively reviewed in recent years and selected examples are shown in Scheme 4 [45, 46].

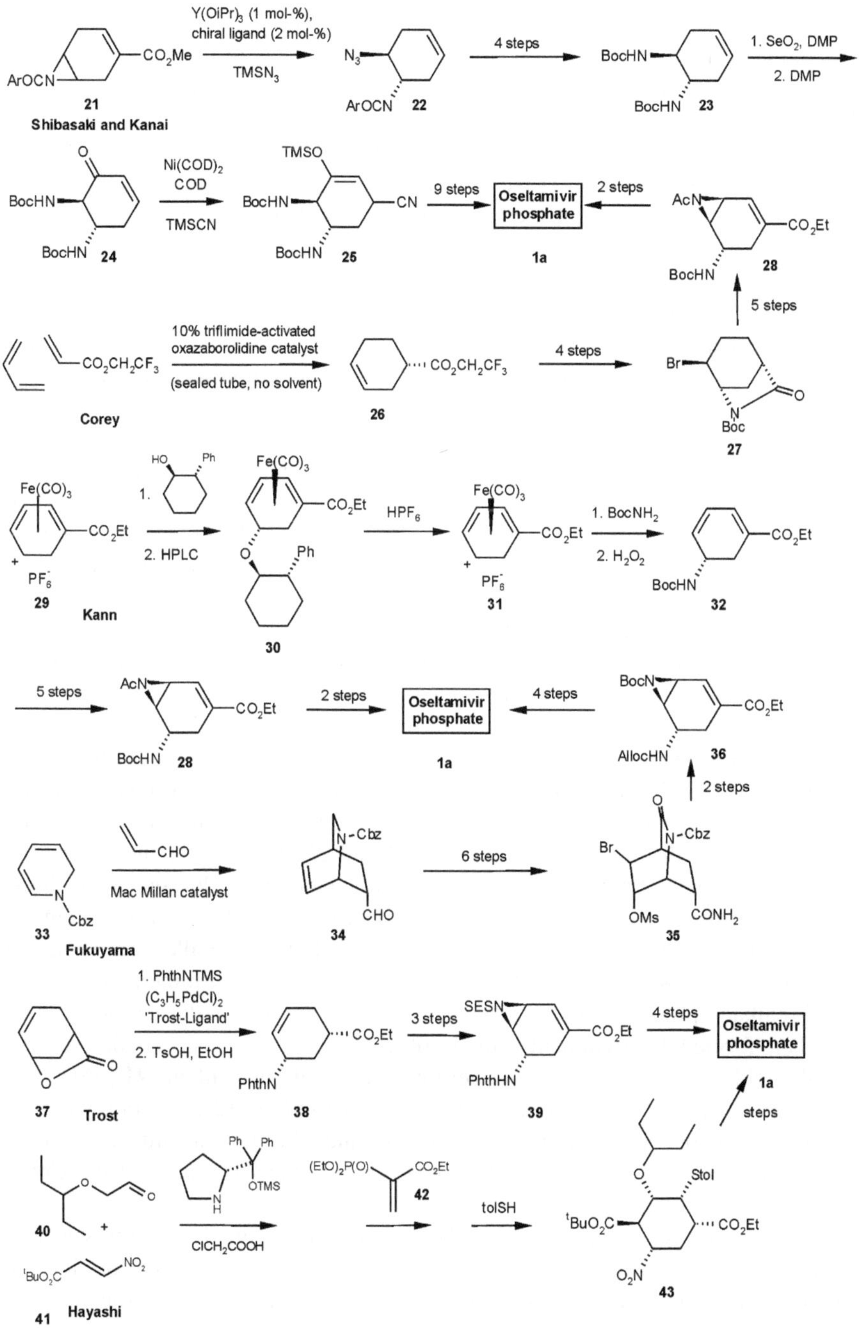

Scheme 4 Selected, chiral pool-independent synthetic approaches to oseltamivir phosphate **1a**

Shibasaki and Kanai synthesised the racemic aziridine **21** in a few steps from 1.4-cyclohexadiene. This underwent ring opening to give **22** in a catalytic, asymmetric fashion with high enantiomeric excess. Conversion to the di-Boc-protected diamine **23** was followed by two oxidation steps in one pot where immediate oxidation of the allyl alcohol to the ketone prevented the over-oxidation of **24** to the allylic dialcohol. An improved version which made some protecting group modifications redundant and did not require toxic SeO_2 was published by the same group as well as the synthesis of a key intermediate via a Diels–Alder strategy [47–49]. Corey's group utilised their catalytic, enantioselective Diels–Alder methodology to obtain the chiral framework **26**, followed by stepwise conversion to the inhibitor with a selective stereoselective bromoamidation to give **27** as the key step [50]. A contribution by Kipassa et al. provided a short synthesis of a precursor of Corey's acetyl aziridine **28** [51].

Kann's laboratory used the readily available, enantiomerically pure iron carbonyl complex **29** to construct the essential chirality, a HPLC separation of **30** was subsequently required but regeneration of the complex ion from the undesired diastereomer was possible [52]. Addition of Boc-amide to **31** followed by oxidative cleavage of the complex gave **32** which was converted into Tamiflu™ **1a**, again via the popular intermediate **28**. Fukuyama's asymmetric Diels–Alder reaction, promoted by the MacMillan catalyst, allowed them to use readily available dihydropyridine derivative **33** as an inexpensive starting material. Adduct **34** was then converted via **35**, into the protected aziridine–carbamate intermediate **36**, typical for these approaches, which allows introduction of the pentyloxy side chain [53]. Further improvements to this synthesis have been made since [54]. Recently, Trost and Zhang reported their short synthesis using their own Pd-ligand system for desymmetrisation of bicyclic lactone **37** at an early stage as the key step with the rest of the transformations, via **38** and **39**, resembling the competing approaches earlier (Scheme 4) [55]. Hayashi's laboratory devised a synthesis consisting of three one-pot operations the first of which is depicted in Scheme 4. Diphenylprolinol silyl ether-catalysed Michael addition of the alkoxyaldehyde **40** to the nitroalkene **41** was followed by a Horner–Wadsworth–Emmons reaction of the nitroalkane product with vinylphosphonate **42**. The major diastereoisomer **43** was then formed by treatment with *p*-toluenethiol under basic conditions. An overall yield of 57% was achieved over nine steps from nitroalkene **41** [56].

Racemic Tamiflu™ has been recently synthesised in 12% overall yield by a cycloaddition between an *N*-protected pyrrole and a bromo-substituted propionic acid ethyl ester. The synthesis eventually leads into the late stages of the Corey synthesis [57, 58].

Researchers at the manufacturer Roche have also been pursuing enantioselective alternatives to their chiral pool-based route. 2,6-Dimethoxyphenol was used as the carbocylic starting material, which, after three modifications, was hydrogenated in a *cis* manner over a Ruthenium-Alox catalyst. The resultant racemate was resolved through an enantioselective ester hydrolysis using pig liver esterase (PLE). Four additional synthetic steps then yielded **1a** in approximately 30% overall yield [59]. Shie et al. [60] generated the cyclohexenecarboxylic acid scaffold and subsequently obtained Tamiflu™ **1a**, together with the phosphonate congener, from xylose via a strategy originating in Fleet's and Vasella's laboratories for shikimic acid [61, 62] and derivatives. This was later extended by our laboratory towards D- and L-*xylo* cyclohexenephosphonates as sialidase inhibitors [63–66]. Other starting materials for chiral pool based syntheses of **1a** which have been described in the last two years are benzene-derived *cis*-dihydrodiols [67–69], glucose [70], D-mannitol [71], and L-methionine [72].

2.2.3 Oseltamivir Derivatives and Isosteres

Given the extensive optimisation of the oseltamivir pharmacophore and the patent coverage, there is naturally a more selective interest in modifying the molecule further, at least in the academic community. The approach introduced by us is based on the recognition that oseltamivir carboxylate **1** and zanamivir **3** (and peramivir **2**) are not only potent influenza sialidase inhibitors, but also simply the strongest low molecular weight, monovalent ligands for the virus [63, 73–75]. In particular, oseltamivir is also highly specific for the influenza virus sialidase, since there is little or no cross-reactivity with other sialic acid-recognising proteins that is known, it could yield materials of formidable affinity and selectivity where its immobilisation and conjugation possible without impairing the pharmacophore. Indeed a recent attempt derived from a total synthesis showed that modification of positions other than the carboxylate position is problematic [76]. It has been long known that phosphonates can function as isosteres of sialic acid carboxylates [77, 78], e.g. in the phospha-isostere of DANA **48** [79] and we have suggested that if the carboxylate could be replaced by phosphonate, the successful oseltamivir motif could be conjugated as a monoester, as these retain a negative charge essential for recognition by the enzyme's arginine triad under physiological conditions [63].

We, and others, introduced a chiral-pool approach from xylose to synthesise such a phospha-oseltamivir [60, 64, 65] and subsequently we showed that indeed little to no penalty in affinity was paid when the oseltamivir pharmacophore was conjugated via phosphonate monoesters to typical selected hydrophobic and hydrophilic structures **46** (e.g. 0.29 μM of **47** against NA(A) H1N1 compared to 0.15 nM for oseltamivir **1** and 0.17 nM for its phospha-isoster **45**, Scheme 5) [73, 74].

The latter examples were synthesised via a Hunsdiecker–Barton halodecarboxylation strategy [75] directly from the industrial oseltamivir precursor, the acetamido-azide **44**.

Dynamic combinatorial libraries have also been exploited to systematically vary the alkylamine glycerol side chain mimetics [80]. Finally, a variety of modifications have been investigated at all positions in the cyclohexene system but these resulted in reduced activity in either enzymatic or plaque reduction assays [6]. The new hydrophobic interaction with S2, which was discovered during the development of the pyrrolidine-based inhibitors in the Abbot laboratories (described below), was also investigated with oseltamivir. Inhibitory potency remained strong, although it was weaker than that of the parent compound [81]. Finally, it should be mentioned here that saturation of the oseltamivir carbocycle, leading to 6-carba-sialic acid and

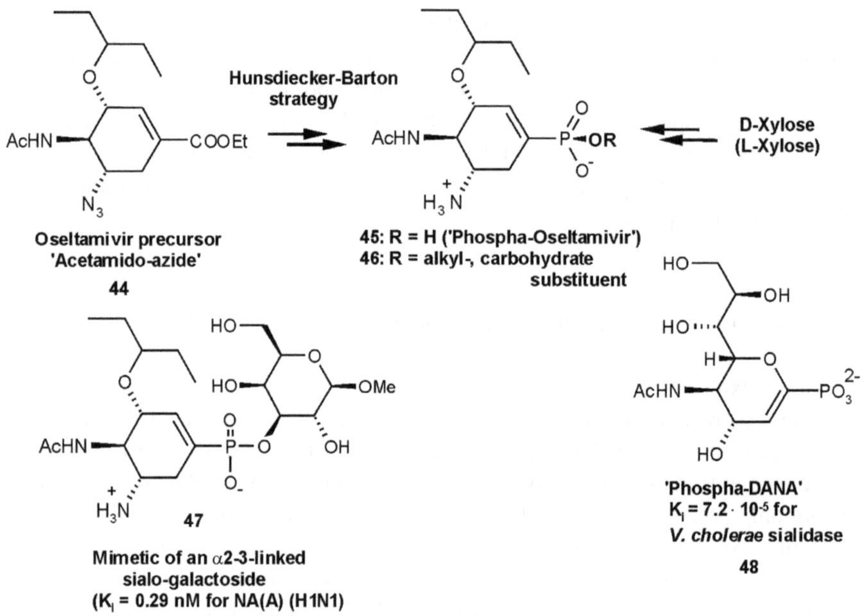

Scheme 5 Phospha-oseltamivir derivatives as a means to conjugate the oseltamivir pharmacophore

sialoside analogues carrying the amine and the pentyloxy side chain displayed much weaker inhibition than their unsaturated counterparts [82].

2.3 *Peramivir and Inhibitors Based on (Non-aromatic) Five-Membered Carbocyclic Scaffolds*

Peramivir **2**, a cyclopentane derivative which is also known as BCX 1812 or RWJ-270201, is the third sialidase inhibitor approved for treatment of influenza so far, albeit with far reaching restrictions. Under the Emergency Use Act, the FDA approved peramivir as a means to 2009 at the height of the swine flu pandemic for emergency administration in an intramuscular and intravenous formulation under the following three conditions: [83] (1) The patient is not responding to either oral (e.g. oseltamivir) or inhaled (e.g. zanamivir) therapy or (2) when drug delivery by a route other than the intravenous route (e.g. enteral or inhaled) is not expected to be dependable or feasible and (3) for adults only, when the clinician judges IV therapy is appropriate due to other patient circumstances. The approval ended when the alert was officially lifted. In Japan, the drug was also approved in January 2010. Peramivir sales 2009 were only about \$22 million compared to oseltamivir's > \$2 billion.

The use of five-membered rings as scaffolds for NA inhibitors (Scheme 6) has started to be more intensely investigated when it was discovered that furanose **49** can inhibit influenza NA as strongly as DANA **4** [84]. The strong binding of **49**

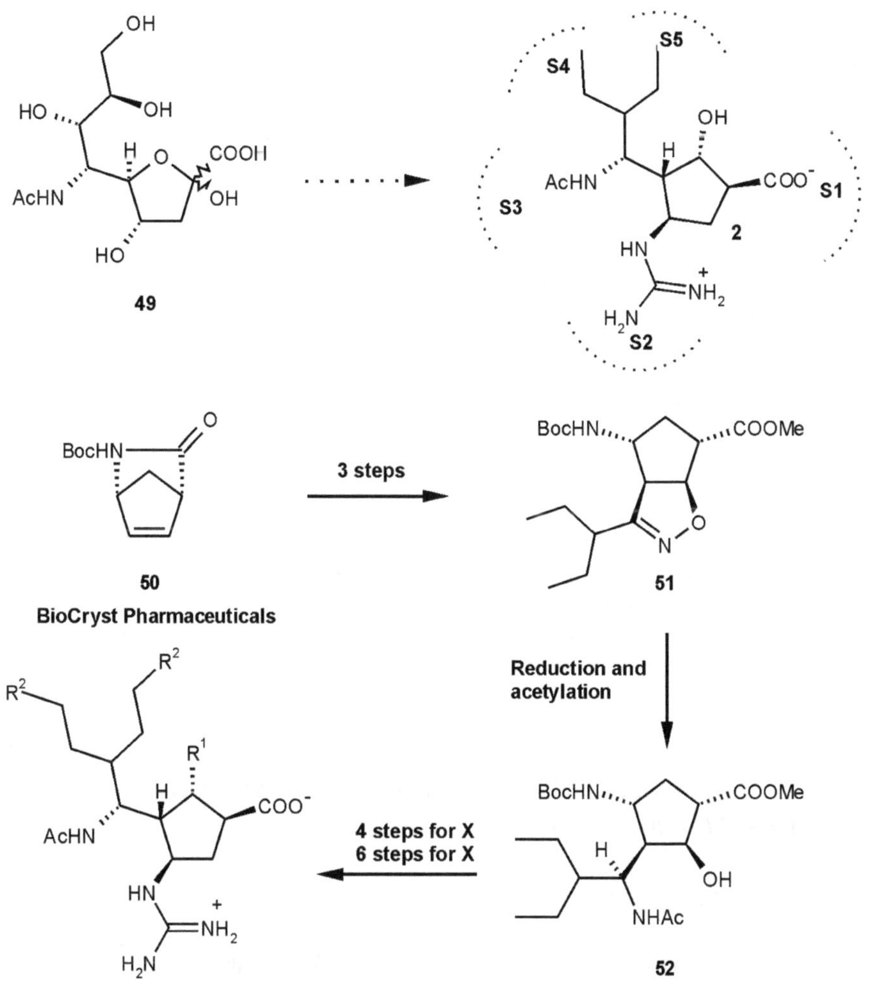

2: R^1 = OH, R^2 = H: Peramivir, BCX-1812
53: R^1 = H: R^2 = H: BCX-1827
54: R^1 = OH, R^2 = CH_3: BCX-1923
55: R^1 = H, R^2 = CH_3: BCX-1898

Scheme 6 Schematic representation of the orientation of peramivir, designed on the basis of furanose **49** as lead compound, in the NA active site and original synthesis of peramivir and derivatives. The longer hydrocarbon chains in **54** and **55** were introduced via the same route

illustrates the fact that it is not the type of scaffold but rather its ability to position the substituents interacting with the various NA subsites in a favourable spatial arrangement that is the crucial point in the inhibition of influenza NA.

With **49** as a lead, the team at BioCryst Pharmaceuticals optimised the interactions through what was a relatively new strategy at that time, relying on co-crystallisation of the enzyme with stereoisomeric mixtures of inhibitors [85].

Repeatedly the more active isomer was identified this way and only after BCX 1812 was identified was a stereoselective synthesis devised [86].

Eventually, four candidates (**2**, **53**–**55**) showed similar or even better efficacy *in vivo* than oseltamivir and zanamivir when given orally or intranasally [87]. Compound **55** (BCX-1898) fell short of the others at lower dose levels and phase III clinical trials with BCX-1812 revealed that the oral bioavailability was insufficient. As **53** and **54** were expected to behave in a similar way due to their similar structure, BCX-1812 remained the final choice as its synthesis was the most efficient (six steps from isoxazoline) [86, 87]. Following methanolysis of lactam **50** and *N*-protection with the Boc-group, a cycloaddition with 2-ethylbutyronitrile was carried out with the isoxazoline **51**. Reduction and *N*-acetylation gave the amide **52** and the carbamate was subsequently converted into the guanidino group. Cleavage of the ester yielded the target BCX 1812 **2** (R = OH) [85, 86]. A variation of this approach produced a synthesis that permitted the facile variation of substituents [88]. The BioCryst team further extended the study of the cyclopentanes to a library of inhibitors with amide side chains and an inverted absolute configuration in the ring and the side chain. Generally, these modifications led to an increase in differential inhibition of NA(A) (preferred) over NA(B) [89].

Diaminocyclitols based on five-membered rings have very recently been suggested as influenza NA inhibitors and a diversity-oriented synthesis was presented but the respective inhibition data have yet to be described [90].

2.4 A-315675 and Inhibitors Based on (Non-aromatic) Five-Membered Heterocyclic Scaffolds

2.4.1 Pyrrolidines as Inhibitors

The use of pyrrolidines as an approach to effective sialidase inhibition was championed mainly by teams at Abbot Laboratories. A lead compound with a submillimolar inhibition of NA was identified from a chemical library and the pyrrolidine scaffold was then systematically investigated using all synthetic, analytical and computational tools of modern drug development [91]. In particular, X-ray data proved to be essential as this series tended to surprise with new novel binding modes [92]. The lead compound **56** did not show significant interaction with subsites S2 and S3 (the amino- and the acetamido pocket of oseltamivir) and therefore the addition of such interactions seemed promising. A plateau with activities in the low micromolar range and with some preference of influenza NA (A) over NA(B) was reached with a series of tetrasubstituted pyrrolidines **57**. However, the carboxylate and hydrophobic substituents interacted as expected (oseltamivir-type) with subsites S1 and S5 respectively and the ionic interaction of the amine with S2, which was crucial for the anticipated several orders of magnitude better inhibition could not be observed (Scheme 7) [92].

An unexpected and novel hydrophobic interaction of the methyl ester in cyclopentane **58** with subsite S2 and the concomitant reorientation of the ring

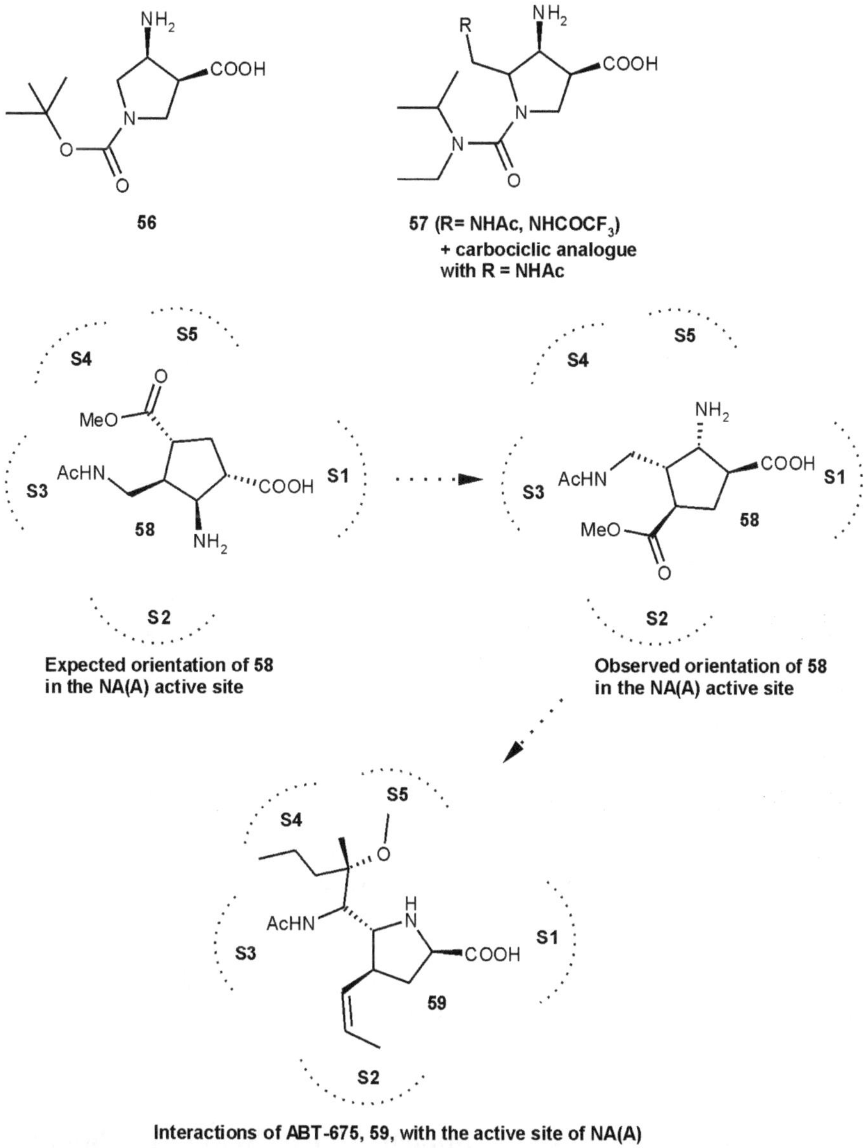

Scheme 7 Schematic representation of important steps in the development of pyrollidine-based inhibitor ABT-675 **59**

which rendered the exocyclic amine virtually obsolete then opened up a new structural space to improve this series (Scheme 7). When additional substituents for hydrophobic interaction with S5 were introduced, the inhibition improved markedly into the subnanomolar range for both influenza NA(A) and (B) [93, 94]. Compound A-315675 (ABT-675, R = H) **59**, which is more active than its carbocyclic or furan-based analogues, shows similar *in vitro* activity to that of zanamivir and oseltamivir against influenza NA(A) and (B). The corresponding ethyl ester prodrug candidate (A-322278, ABT-667) was as effective as the established inhibitors in reducing viral replication, reducing weight loss and prolonging survival in an immunocompromised murine model of an influenza infection. Another interesting property is the lack of cross-resistance towards oseltamivir and A-315675 when tested with oseltamivir-resistant influenza sialidases of N1 and N2 subtypes [95].

The Abbott group and others [96] have continued to study pyrrolidines. The earlier work by Abbott mainly examined variation of the S5 interaction of A-315675. Comparable activity (for influenza NA(A) and (B) or (B) alone) has been reported [97]. Finally, A-315675 and derivatives are α-amino acids and it has to be noted here that a study with related β-amino acids, i.e. a shift of the ring nitrogen position with respect to the carboxylate, has also been carried out [98].

2.4.2 Syntheses of A-315675

The initial synthesis of A-315675 **59** was not diastereoselective and the overall yield was only moderate. This prompted competition for a more effective synthesis, albeit not as extensive as in the oseltamivir case. The Abbott group devised a stereoselective route to A-315675 giving access to multigram quantities [99] which was later improved [100] to yield kilogram quantities of the isopropyl ester as a prodrug for pre-clinical evaluation. Core steps are a Mannich condensation of **61** and a pyrrole derivative to give **62** which was then subjected to a cuprate-promoted Michael addition. Product **63** was then converted into the inhibitor **59** in six steps (Scheme 8) [99, 100].

Hanessian and co-workers' highly stereocontrolled synthesis gave A-315675 **59** in 22 steps and 12.8% overall yield from D-serine [101]. Crucial steps in the heterocycle formation are the highly diastereoselective formation of **65** by the addition of ethyl propiolate to nitrone **64** and the subsequent cyclisation to lactam **66**. Protecting group differences aside, the final steps, including the Michael addition to furnish **67**, then correspond to those of the Abbott synthesis.

3 Non-carbohydrate-Based Influenza Virus Sialidase Inhibitors Structurally Unrelated to Sialic Acids

With interest in novel lead structures for the development of new generations of influenza NA inhibitors constantly growing, it is not surprising that screening for anti-NA activity is now more often being included in the screening of natural products and synthetic libraries. In Fig. 5, a few examples of such structures are depicted. Coumarin **68** showed the highest *in vitro* activity ($IC_{50} = 3.1$ μM against influenza NA(A), non-competitive inhibition) in a study of polyphenols isolated

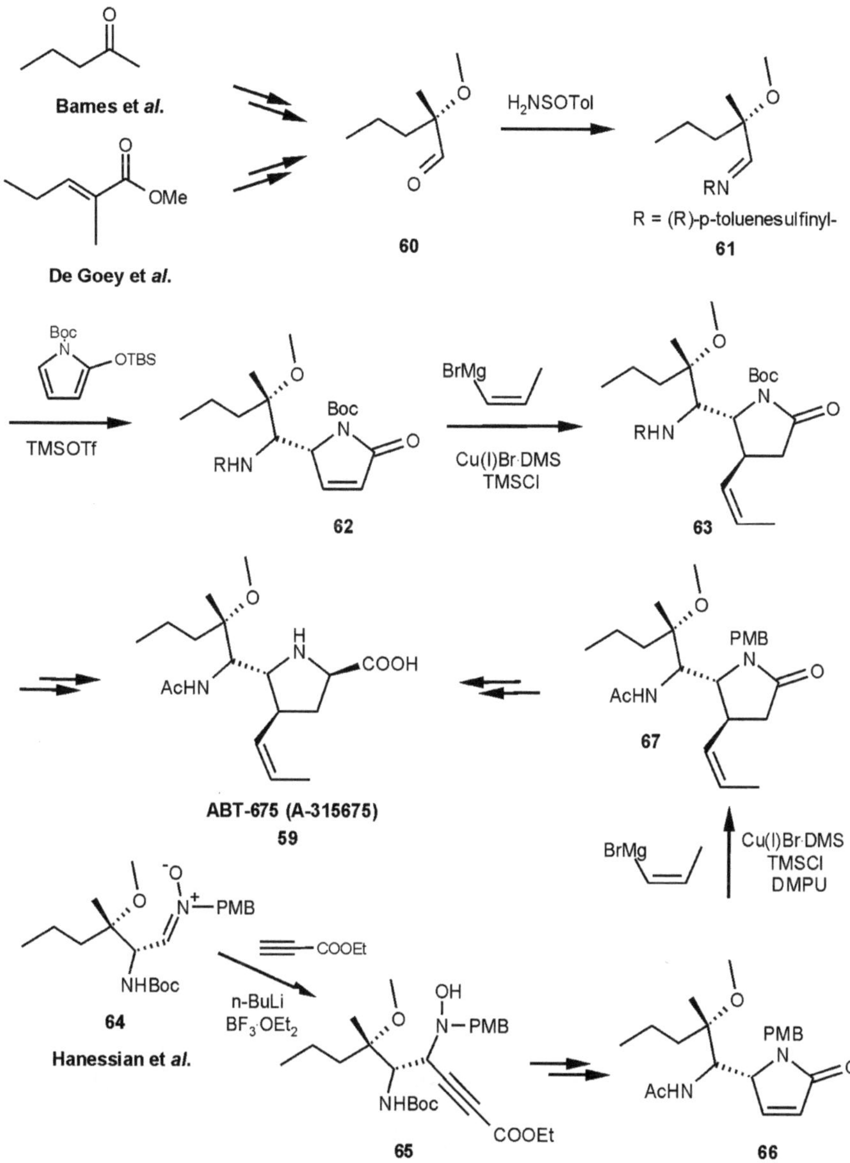

Scheme 8 Syntheses of pyrrolidine-based inhibitor A-315675 **59**

from *Glycyrrhiza uralensis* roots [102]. Kaempferol **69**, a flavonoid, from *Rhodiola rosea* roots, was active against NA(A) ($IC_{50} = 8.1$ μM, non-competitive) and also against the virus in a cytopathic effect inhibition (CPE) assay [103].

Compound **70** is another, non-competitive submicromolar inhibitor of NA(A) from *Cudrania tricuspidata* [104]. With substantial numbers of these phenol-type

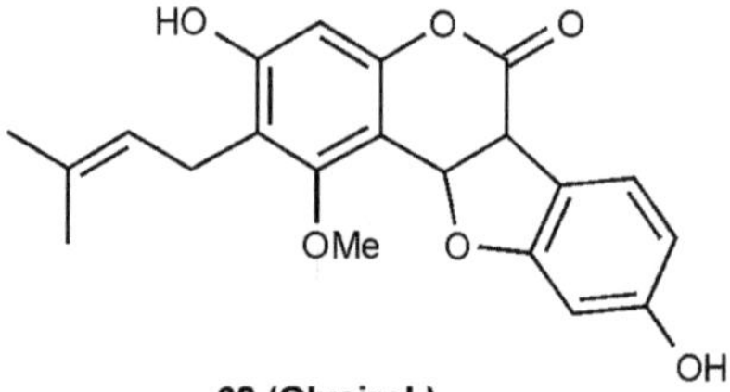

68 (Glycirol)

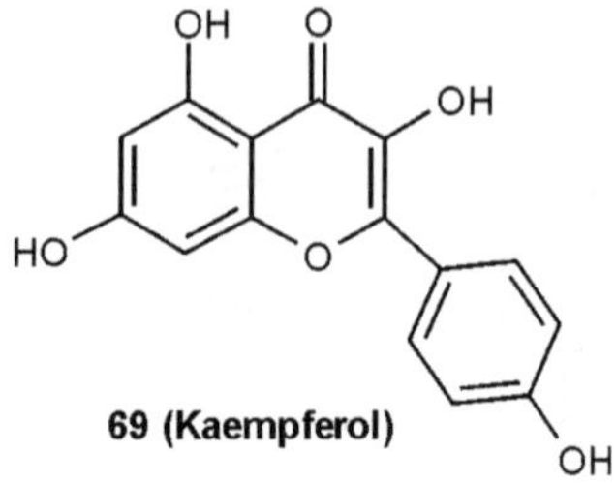

69 (Kaempferol)

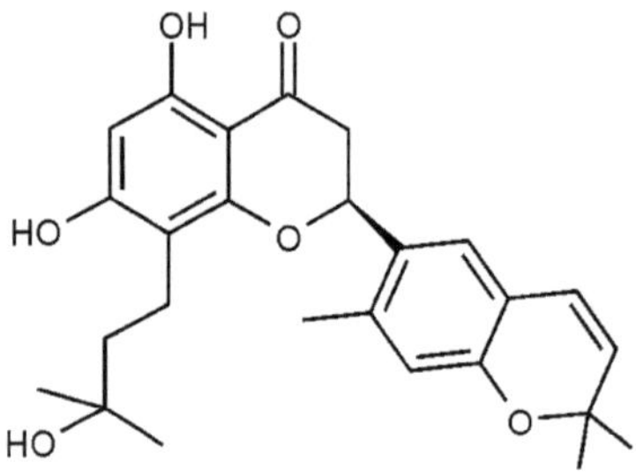

70

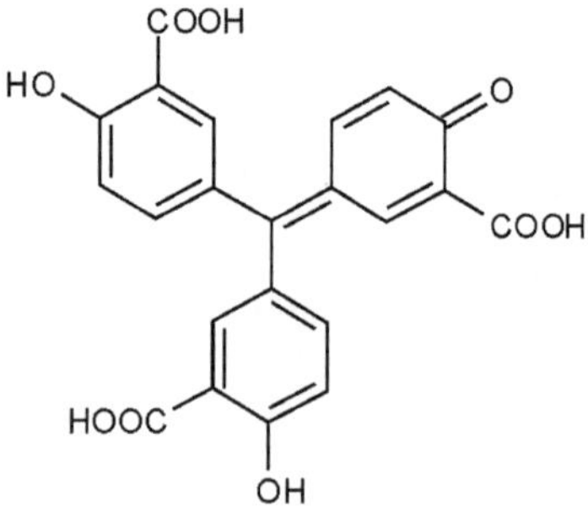

71 (ATA)

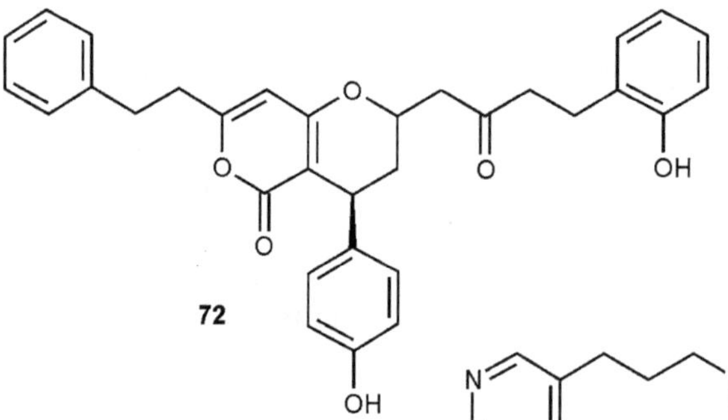

72

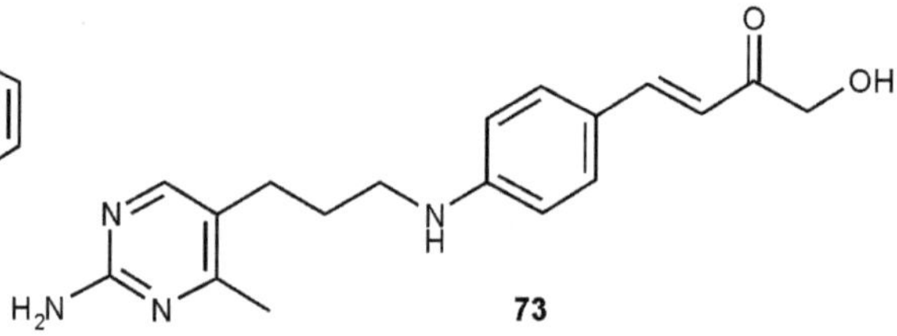

73

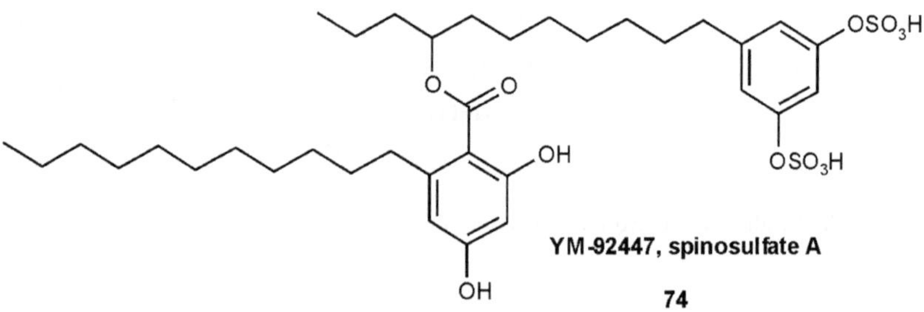

YM-92447, spinosulfate A

74

Fig. 5 Examples of natural products and compounds from synthetic libraries exhibiting influenza NA inhibitory activity

inhibitors with varying activity now available, it can be expected that a full understanding of their SAR with influenza NA is not far away [104, 105].

Library screening revealed that aurintricarboxylic acid (ATA) **71** inhibits influenza NA(A) (wild type and mutant H274Y) in the low micromolar range. Full kinetic data are not yet available but a modelling/docking study suggested that it indeed binds to the influenza NA active site. It also inhibited virus replication [106].

Diarylheptanoids like katsadain A **72** from *Alpinia katsadai* are another group of natural products with low micromolar activity against influenza NA(A) and a suspected interaction with the enzyme active site [107].

As nature is obviously not the only source of compound libraries with potential influenza NA inhibitory activity, computational screening of synthetic libraries represents another promising approach. This is further enhanced as increasingly detailed information about the structure of sialidase subgroups, such as the open conformation of group 1 sialidases, becomes available, thus expanding the structural space available for binding. In such a study, 230,000 compounds from the National Cancer Institute (NCI), USA database were screened computationally for binding to this open-form binding site. Compound **73** was selected and investigated further. It showed relatively weak activity, only at a 100 μM concentration could inhibition be observed, but the effect seemed to be additive when oseltamivir was present (1 μM), thus suggesting a different mode of binding [108].

Spinosulfate A **74**, which was isolated from a pycnidial fungus, exhibited an IC_{50} (0.5 μM, mixed inhibition) comparable to DANA against an influenza NA(A). Kinetic analysis suggested mixed inhibition and cytotoxicity was found to be about 100-fold weaker than sialidase activity [109].

4 *In-Silico*-Generated Potential Influenza Virus Sialidase Inhibitors

There is a growing number of studies in the literature that report computational design of potential new influenza NA inhibitors, without the compounds being synthesised or tested with NA. Although some studies are very thorough in calculating the potential free energy of binding, taking oseltamivir or zanamivir as a reference point, they have to be viewed with great caution. There is no doubt that, for instance, the '150-cavity' of group 1 sialidases or sialidases of oseltamivir-resistant viruses offer opportunities to introduce additional interactions to the oseltamivir structure *in silico*, in particular once X-ray data become available [13, 110–112]. However, as described earlier, experience tells us that there are events that cannot easily be predicted. Thus, the observation that the hydrophobic side chains that are so important in the oseltamivir series, can be accommodated by the enzyme through a flip of a hydrophilic glutamate, was unexpected. Likewise, the flipping of the pyrrolidine ring during the development of A-315675 came as a surprise. Therefore, as long as there is no hard evidence, for instance through an enzyme essay, that a molecule indeed inhibits sialoside hydrolysis by influenza NA, it is at the best a 'potential' inhibitor [113, 114].

Acknowledgments The authors thank Prof. Jim Hanson for proofreading the manuscript and helpful advice.

References

1. von Itzstein M (2007) The war against influenza: discovery and development of sialidase inhibitors. Nat Rev Drug Discov 6:967–974
2. De Clerq E (2006) Antiviral agents active against influenza A viruses. Nat Rev 5:1015–1025
3. Meindl P, Bodo G, Palese P, Schulman J, Tuppy H (1974) Inhibition of neuraminidase activity by derivatives of DANA. Virology 58:457–463
4. Burmeister W, Henrissat B, Bosso C, Cusack S, Ruigrok RWH (1993) Influenza B virus neuraminidase can synthesize its own inhibitor. Structure 1:19–26
5. von Itzstein M, Wu W-Y, Kok GB, Pegg MS, Dyason JC, Jin B, Phan TV, Smythe ML, White HF, Oliver SW, Colman PM, Varghese JN, Ryan DM, Woods JM, Bethell RC, Hotham VJ, Cameron JM, Penn CR (1993) Rational design of potent sialidase-based inhibitors of influenza virus replication. Nature 363:418–423
6. Rich J, Gehle D, von Itzstein M (eds) (2007) Design and synthesis of sialidase inhibitors for influenza virus infections. Elsevier, Amsterdam
7. Kiefel M, von Itzstein M (2002) Recent advances in the synthesis of sialic acid derivatives and sialylmimetics as biological probes. Chem Rev 102:471–490
8. Dyason JC, von Itzstein M (2001) Anti-influenza virus drug design: sialidase inhibitors. Aust J Chem 54:663–670
9. Streicher H (2002) Approaches to carboyclic sialidase inhibitors. Monatsh Chem Chem Month 133:1263–1278
10. Busse H (ed) (2007) Recent developments in the synthesis and application of sialylimetics. Scion Publishing, Bloxham
11. Liu Y, Zhang J, Xu W (2007) Recent progress in rational drug design of neuraminidase inhibitors. Curr Med Chem 14:2872–2891
12. Angata T, Varki A (2002) Chemical diversity in the sialic acids. Chem Rev 102:439–469
13. Russell RJ, Haire LF, Stevens DJ, Collins PJ, Lin YP, Blackburn GM, Hay AJ, Gamblin SJ, Skehel JJ (2006) The structure of H5N1 avian influenza neuraminidase suggests new opportunities for drug design. Nature 443:45–49
14. Williams M, Bischofberger N, Swaminathan S, Kim CU (1995) Synthesis and influenza neuraminidase inhibitory activity of aromatic analogues of sialic acid. Bioorg Med Chem Lett 5:2251–2254
15. Jedrzejas M, Singh S, Brouillette WJ, Laver WG, Air GM, Luo M (1995) Structures of aromatic inhibitors of influenza virus neuraminidase. Biochemistry 34:3144–3151
16. Chand P, Babu YS, Bantia S, Chu N, Cole LB, Kotian PL, Laver WG, Montgomery JA, Pathak VP, Petty SL, Shrout DP, Walsh DA, Walsh GM (1997) Design and synthesis of benzoic acid derivatives as influenza neuraminidase inhibitors using structure-based drug design. J Med Chem 40:4030–4052
17. Singh SJM, Air GM, Luo M, Laver WG, Brouilette WJ (1995) Structure-based inhibitors of influenza virus sialidase. A benzoic acid lead with novel interaction. J Med Chem 38:3217–3225
18. Chand P, Kotian PL, Morris PE, Bantia S, Walsh DA, Babu YS (2005) Synthesis and inhibitory activity of benzoic acid and pyridine derivatives on influenza neuraminidase. Bioorg Med Chem 13:2665–2678

19. Sudbeck E, Jedrzejas MJ, Singh S, Brouilette WJ, Air GM, Laver WG, Babu YS, Bantia S, Chand P, Chu N, Montgomery JA, Walsh DA, Luo M (1997) Guanidinobenzoic acid inhibitors of influenza virus neuraminidase. J Mol Biol 267:584–594
20. Howes PD, Cleasby A, Evans DN, Feilden H, Smith PW, Sollis SL, Taylor N, Wonacott AJ (1999) 4-Acetylamino-3-(imidazol-1-yl)-benzoic acids as novel inhibitors of influenza sialidase. Eur J Med Chem 34:225–234
21. Brouillette W, Atigadda VR, Luo M, Air GM, Babu YS, Bantia S (1999) Design of benzoic acid inhibitors of influenza neuraminidase containing a cyclic substitution for the N-acetyl grouping. Bioorg Med Chem Lett 9:1901–1906
22. Finley JB, Atigadda VR, Duarte F, Zhao JJ, Brouillette WJ, Air GM, Luo M (1999) Novel aromatic inhibitors of influenza virus neuraminidase make selective interactions with conserved residues and water molecules in the active site. J Mol Biol 293:1107–1119
23. Brouillette WJ, Bajpai SN, Ali SM, Velu SE, Atigadda VR, Lommer BS, Finley JB, Luo M, Air GM (2003) Pyrrolidinobenzoic acid inhibitors of influenza virus neuraminidase: modifications of essential pyrrolidinone ring substituents. Bioorg Med Chem 11:2739–2749
24. Atigadda VR, Brouillette W, Duarte F, Ali SM, Babu YS, Bantia S, Chand P, Chu N, Montgomery JA, Walsh DA, Sudbeck EA et al (1999) Potent inhibition of influenza sialidase by a benzoic acid containing a 2-pyrrolidinone substituent. J Med Chem 42:2332–2343
25. Zhang J, Wang Q, Fang H, Xu W, Liu A, Du G (2008) Design, synthesis, inhibitory activity, and SAR studies of hydrophobic p-aminosalicylic acid derivatives as neuraminidase inhibitors. Bioorg Med Chem 16:3839–3847
26. Flashner M, Kessler J, Tanenbaum SW (1983) The interaction of substrate-related ketals with bacterial and viral neuraminidases. Arch Biochem Biophys 221:188–196
27. Taylor NR, Cleasby A, Singh O, Skarzynski T, Wonacott AJ, Smith PW, Sollis SL, Howes PD, Cherry PC, Bethell R et al (1998) Dihydropyrancarboxamides related to zanamivir: a new series of inhibitors of influenza virus sialidases. 2. Crystallographic and molecular modeling study of complexes of 4-amino-4H-pyran-6-carboxamides and sialidase from influenza virus types A and B. J Med Chem 41:798–807
28. Smith PW, Sollis SL, Howes PD, Cherry PC, Starkey ID, Cobley KN, Weston H, Scicinski J, Merritt A, Whittington A et al (1998) Dihydropyrancarboxamides related to zanamivir: a new series of inhibitors of influenza virus sialidases. 1. Discovery, synthesis, biological activity, and structure and activity relationships of 4-guanidino- and 4-amino-4H-pyran-6-carboxamides. J Med Chem 41:787–797
29. Wyatt PG, Coomber BA, Evans DN, Jack TI, Fulton HE, Wonacott AJ, Colman P, Varghese J (2001) Sialidase inhibitors related to zanamivir. Further SAR studies of 4-amino-4H-pyran-2-carboxylic acid-6-propylamides. Bioorg Med Chem Lett 11:669–673
30. Kerrigan SA, Smith PW, Stoodley RJ (2001) Synthesis of (4R*,5R*)-4-acetylamino-5-diethylcarbamoylcyclohex-1-ene-1-carboxylic acid and (3R*,4R*)-4-acetylamino-3-diethylcarbamoylcyclohex-1-ene-1-carboxylic acid: new inhibitors of influenza virus sialidases. Tetrahedron Lett 42:4709–4712
31. Taylor NR, von Itzstein M (1994) Molecular modeling studies on ligand binding to sialidase from influenza virus and the mechanism of catalysis. J Med Chem 37:616–624
32. Vorwerk S, Vasella A (1998) Carbocyclische analoga von NANA. Angew Chem Int Ed Engl 37:1732–1734
33. Chandler M, Conroy R, Cooper AWJ, Lamont RB, Scicinski JJ, Smart JE, Storer R, Weir NG, Wilson RD, Wyatt PG (1995) Approaches to carbocyclic analogues of the potent neuraminidase Inhibitor 4-guanidino-Neu5Ac2en. X.Ray molecular structure of N-[(1 S,2S,6R)-2-Azido-6-benzyloxymethyl-4-formylcyclohex-3-enyl]acetamide. J Chem Soc Perkin Trans 1:1189–1197
34. Kim CU, Lew W, Williams MA, Liu H, Zhang L, Swaminathan S, Bischofberger N, Chen MS, Mendel DB, Tai CY, Laver WG, Stevens RC (1997) Structure – activity relationship studies of novel carbocyclic influenza neuraminidase inhibitors. J Am Chem Soc 119:681–690
35. Kim C, Lew W, Williams MA, Zhang L, Chen X, Escarpe PA, Mendel DB, Laver WG, Stevens RCJ (1998) Influenza neuraminidase inhibitors possessing a novel hydrophobic

interaction in the enzyme active site: design, synthesis, and structural analysis of carbocyclic sialic acid analogues with potent anti-influenza activity. J Med Chem 41:2451–2460
36. Williams MA, Lew W, Mendel DB, Tai CY, Escarpe PA, Laver WG, Stevens RC, Kim CU (1997) Structure-activity relationships of carbocyclic influenza neuraminidase inhibitors. Bioorg Med Chem Lett 7:1837–1842
37. Lew W, Wu H, Mendel DB, Escarpe PA, Chen X, Laver WG, Graves BJ, Kim CU (1998) A new series of C3-aza carbocyclic influenza neuraminidase inhibitors: synthesis and inhibitory activity. Bioorg Med Chem Lett 8:3321–3324
38. Lew W, Wu H, Chen X, Graves BJ, Escarpe PA, MacArthur HL, Mendel DB, Kim CU (2000) Carbocyclic influenza neuraminidase inhibitors possessing a C3-cyclic amine side chain: synthesis and inhibitory activity. Bioorg Med Chem Lett 10:1257–1260
39. Lew W, Williams MA, Mendel DB, Escarpe PA, Kim CU (1997) C3-Thia and C3-carba isosteres of a carbocyclic influenza neuraminidase inhibitor, (3R,4R,5 S)-4-acetamido-5-amino-3-propoxyl-1-cyclohexene-1-carboxylic acid. Bioorg Med Chem Lett 7:1843–1846
40. Zhang L, Williams MA, Mendel DB, Escarpe PA, Kim CU (1997) Synthesis and activity of C2-substituted analogs of influenza neuraminidase inhibitor GS 4071. Bioorg Med Chem Lett 7:1847–1850
41. Rohloff JC, Kent KM, Postich MJ, Becker MW, Chapman HH, Kelly DE, Lew W, Louie MS, McGee LR, Prisbe EJ et al (1998) Practical total synthesis of the anti-influenza drug GS-4104. J Org Chem 63:4545–4550
42. Abrecht S, Harrington P, Iding H, Karpf M, Trussardi R, Wirz B, Zutter U (2004) The synthetic development of the anti-influenza neuraminidase inhibitor oseltamivir phosphate (tamiflu): a challenge for synthesis and process research. Chimia 58:621–629
43. Abrecht S, Federspiel MC, Estermann H, Fischer R, Karpf M, Mair H-J, Oberhauser T, Rimmler G, Trussardi R, Zutter U (2007) The synthetic-technical development of oseltamivir phosphate tamiflu^{tm}: a race against time. Chimia 61:93–99
44. Karpf M, Trussardi R (2009) Effizienter zugang zu oseltamivirphosphat (tamiflu) über das O-trimesylat von shikimisäureethylester. Angew Chem Int Ed Engl 121:5871–5873
45. Magano J (2009) Synthetic approaches to the neuraminidase inhibitors zanamivir (relenza) and oseltamivir phosphate (tamiflu) for the treatment of influenza. Chem Rev 109:4398–4438
46. Shibasaki M, Kanai M (2008) Synthetic strategies for oseltamivir phosphate. Eur J Org Chem:1839–1850
47. Fukuta Y, Mita T, Fukuda N, Kanai M, Shibasaki M (2006) De novo synthesis of tamiflu via a catalytic asymmetric ring-opening of meso-aziridines with TMSN3. J Am Chem Soc 128: 6312–6313
48. Mita T, Fukuda N, Roca FX, Kanai M, Shibasaki M (2007) Second generation catalytic asymmetric synthesis of tamiflu: allylic substitution route. Org Lett 9:259–262
49. Yamatsugu K, Kamijo S, Suto Y, Kanai M, Shibasaki M (2007) A concise synthesis of Tamiflu: third generation route via the Diels–Alder reaction and the Curtius rearrangement. Tetrahedron Lett 48:1403–1406
50. Yeung Y-Y, Hong S, Corey EJ (2006) A short enantioselective pathway for the synthesis of the anti-influenza neuraminidase inhibitor oseltamivir from 1,3-butadiene and acrylic acid. J Am Chem Soc 128:6310–6311
51. Kipassa N, Okamura H, Kina K, Hamada T, Iwagawa T (2008) Efficient short step synthesis of Corey's tamiflu intermediate. Org Lett 10:815–816
52. Bromfield K, Graden H, Hagberg DP, Olsson T, Kann N (2007) An iron carbonyl approach to the influenza neuraminidase inhibitor oseltamivir. Chem Commun: 3183–3185
53. Satoh N, Akiba T, Yokoshima S, Fukuyama T (2007) A practical synthesis of (−) oseltamivir. Angew Chem Int Ed Engl 46:5734–5736
54. Satoh N, Akiba T, Yokoshima S, Fukuyama T (2009) A practical synthesis of (−)-oseltamivir. Tetrahedron 65:3239–3245
55. Trost BM, Zhang T (2008) A concise synthesis of (−) oseltamivir. Angew Chem 120: 3819–3821

56. Ishikawa H, Suzuki T, Hayashi Y (2009) High-yielding synthesis of the anti-influenza neuramidase inhibitor (−)-oseltamivir by three ln-one-pot-operations. Angew Chem Int Ed Engl 48:1304–1307
57. Sun H, Lin YJ, Wu YL, Wu YK (2009) A facile access to antiflu agent tamiflu/oseltamivir. Synlett: 2473–2476
58. Kamimura A, Nakano T (2010) Use of the diels-alder adduct of pyrrole in organic synthesis. Formal racemic synthesis of tamiflu. J Org Chem 75:3133–3136
59. Zutter U, Iding H, Spurr P, Wirz B (2008) New, efficient synthesis of oseltamivir phosphate (tamiflu) via enzymatic desymmetrization of a meso-1,3-cyclohexanedicarboxylic acid diester. J Org Chem 73:4895–4902
60. Shie J-J, Fang J-M, Wang S-Y, Tsai K-C, Cheng Y-SE, Yang A-S, Hsiao S-C, Su C-Y, Wong C-H (2007) Synthesis of tamiflu and its phosphonate congeners possessing potent anti-influenza activity. J Am Chem Soc: 11893–11894
61. Fleet G, Shing TKM, Warr SM (1984) Enantiospecific synthesis of shikimic acid from d-mannose: formation of a chiralcyclohexene by intramolecular olefination of a carbohydrate-derived intermediate. J Chem Soc Perkin Trans 1:905–908
62. Mirza S, Vasella A (1984) Synthesis of methyl shikimate and of diethyl phosphashikimate from d-ribose. Helv Chim Acta 67:1562–1567
63. Busse H, Streicher H (2006) Building a successful structural motif into sialylmimetics-cyclohexenephosphonate monoesters as pseudo-sialosides with promising inhibitory properties. Bioorg Med Chem 14:1047–1057
64. Streicher H, Meisch J, Bohner C (2001) Synthesis of xylo-configured cyclohexenephosphonates, versatile precursors of sialidase inhibtor libraries. Tetrahedron 57:8851–8859
65. Streicher H, Bohner C (2002) Synthesis of functionalized cyclohexenephosphonates and their inhibitors activity. Tetrahedron 58:7573–7581
66. Streicher H (2004) Synthesis and evaluation as sialidase inhibitors of xylo-configured cyclohexenephosphonates carrying glycerol sidechain mimics. Bioorg Med Chem Lett 14: 361–364
67. Sullivan B, Carrera I, Drouin M, Hudlicky T (2009) Symmetry-based design for the chemoenzymatic synthesis of oseltamivir (tamiflu) from ethyl benzoate13. Angew Chem Int Ed Engl 48:4229–4231
68. Matveenko M, Willis AC, Banwell MG (2008) A chemoenzymatic synthesis of the anti-influenza agent Tamiflu®. Tetrahedron Lett 49:7018–7020
69. Shie J-J, Fang J-M, Wong C-H (2008) A concise and flexible synthesis of the potent anti-influenza agents tamiflu and tamiphosphor13. Angew Chem Int Ed Engl 47:5788–5791
70. Ma J, Zhao Y, Ng S, Zhang J, Zeng J, Than A, Chen P, Liu X-W (2010) Sugar-based synthesis of tamiflu and its inhibitory effects on cell secretion. Chem Eur J 16:4533–4540
71. Mandai T, Oshitari T (2009) Efficient asymmetric synthesis of oseltamivir from d-mannitol. Synlett 2009:783–786
72. Oshitari T, Mandai T (2009) Azide-free synthesis of oseltamivir from l-methionine. Synlett 2009:787–789
73. Carbain BSRM, Collins PJ, Hitchcock PB, Streicher H (2009) Galactose-conjugates of the oseltamivir pharmacophore – new tools for the characterization of influenza virus neuraminidases. Org. Biomol, Chem, 9
74. Carbain B, Collins Patrick J, Callum L, Martin Stephen R, Hay Alan J, McCauley J, Streicher H (2009) Efficient synthesis of highly active phospha-isosteres of the influenza neuraminidase inhibitor oseltamivir. ChemMedChem 4:335–337
75. Carbain B, Hitchcock PB, Streicher H (2010) New aspects of the Hunsdiecker-Barton halodecarboxylation–syntheses of phospha-shikimic acid and derivatives. Tetrahedron Lett 51:2717–2719
76. Kimura Y, Yamatsugu K, Kanai M, Echigo N, Kuzuhara T, Shibasaki M (2009) Design and synthesis of immobilized Tamiflu analog on resin for affinity chromatography. Tetrahedron Lett 50:3205–3208
77. Chan T-H, Xin Y-C, von Itzstein M (1997) Synthesis of phosphonic acid analogues of sialic acids (Neu5Ac and KDN) as potential sialidase inhibitors. J Org Chem 62:3500–3504

78. Wallimann K, Vasella A (1990) Phosphonic acid analogues of the N-Acetyl-2-deoxyneuraminic acids. Helv Chim Acta 73:1359–1372
79. Vasella A, Wyler R (1991) Synthesis of a phosphonic acid analogue of DANA. Helv Chim Acta 74:451–463
80. Hochgurtel M, Biesinger R, Kroth H, Piecha D, Hofmann MW, Krause S, Schaaf O, Nicolau C, Eliseev AV (2003) Ketones as building blocks for dynamic combinatorial libraries: highly active neuraminidase inhibitors generated via selection pressure of the biological target. J Med Chem 46:356–358
81. Hanessian S, Wang J, Montgomery D, Stoll V, Stewart KD, Kati W, Maring C, Kempf D, Hutchins C, Laver WG (2002) Design, synthesis, and neuraminidase inhibitory activity of GS-4071 analogues that utilize a novel hydrophobic paradigm. Bioorg Med Chem Lett 12: 3425–3429
82. Jeong JW, Kim JK, Woo BY, Song BJ, Ha D-C, No Z (2004) A substrate mimetic approach for influenza neuraminidase inhibitors. Bull Korean Chem Soc 25:1575–1577
83. Reuters (2009) FDA authorizes emergency use of intravenous antiviral peramivir for 2009 H1N1 influenza for certain patients, settings. http://www.reuters.com/article/idUS10036+24-Oct-2009+PRN20091024
84. Yamamoto T, Kumazawa H, Inami K, Teshima T, Shiba T (1992) Syntheses of sialic acid isomers with inhibitory activity against neuraminidase. Tetrahedron Lett 33:5791–5794
85. Babu YS, Chand P, Bantia S, Kotian P, Dehghani A, El Kattan Y, Lin T, Hutchison TL, Elliott AJ, Parker CD et al (2000) BCX-1812, a highly potent and selective influenza neuraminidase inhibitor. J Med Chem 43:3482–3486
86. Chand P, Kotian PL, Dehghani A, El-Kattan Y, Lin T-H, Hutchison TL, Babu YS, Bantia S, Elliott AJ, Montgomery JA (2001) Systematic structure-based design and stereoselective synthesis of novel multisubstituted cyclopentane derivatives with potent antiinfluenza activity. J Med Chem 44:4379–4392
87. Chand P, Bantia S, Kotian PL, El-Kattan Y, Lin T-H, Babu YS (2005) Comparison of the anti-influenza virus activity of cyclopentane derivatives with oseltamivir and zanamivir in vivo. Bioorg Med Chem 13:4071–4077
88. Mineno T, Miller MJ (2003) Stereoselective total synthesis of racemic BCX-1812 (RWJ-270201) for the development of neuraminidase inhibitors as anti-influenza agents. J Org Chem 68:6591–6596
89. Chand P, Babu YS, Bantia S, Rowland S, Dehghani A, Kotian PL, Hutchison TL, Ali S, Brouillette W, El-Kattan Y et al (2004) Syntheses and neuraminidase inhibitory activity of multisubstituted cyclopentane amide derivatives. J Med Chem 47:1919–1929
90. Cui Y, Jiao Z, Gong J, Yu Q, Zheng X, Quan J, Luo M, Yang Z (2009) Development of new stereodiverse diaminocyclitols as inhibitors of influenza virus neuraminidase. Org Lett 12: 4–7
91. Wang GT, Chen Y, Wang S, Gentles R, Sowin T, Kati W, Muchmore S, Giranda V, Stewart K, Sham H et al (2001) Design, synthesis, and structural analysis of influenza neuraminidase inhibitors containing pyrrolidine cores. J Med Chem 44:1192–1201
92. Stoll V, Stewart KD, Maring CJ, Muchmore S, Giranda V, Gu Y-GY, Wang G, Chen Y, Sun M, Zhao C et al (2003) Influenza neuraminidase inhibitors: Structure-based design of a novel inhibitor series. Biochemistry 42:718–727
93. Wang GT, Wang S, Gentles R, Sowin T, Maring CJ, Kempf DJ, Kati WM, Stoll V, Stewart KD, Laver G (2005) Design, synthesis, and structural analysis of inhibitors of influenza neuraminidase containing a 2,3-disubstituted tetrahydrofuran-5-carboxylic acid core. Bioorg Med Chem Lett 15:125–128
94. Maring CJ, Stoll VS, Zhao C, Sun M, Krueger AC, Stewart KD, Madigan DL, Kati WM, Xu Y, Carrick RJ et al (2005) Structure-based characterization and optimization of novel hydrophobic binding interactions in a series of pyrrolidine influenza neuraminidase inhibitors. J Med Chem 48:3980–3990
95. Abed Y, Nehmé B, Baz M, Boivin G (2008) Activity of the neuraminidase inhibitor A-315675 against oseltamivir-resistant influenza neuraminidases of N1 and N2 subtypes. Antiviral Res 77:163–166

96. Zhang JXW, Liu A, Du G (2008) Design, synthesis and preliminary evaluation of new pyrrolidine derivatives as neuraminidase inhibitors. Med Chem 4:204–209
97. Krueger AC, Xu Y, Kati WM, Kempf DJ, Maring CJ, McDaniel KF, Molla A, Montgomery D, Kohlbrenner WE (2008) Synthesis of potent pyrrolidine influenza neuraminidase inhibitors. Bioorg Med Chem Lett 18:1692–1695
98. Kati WM, Montgomery D, Maring C, Stoll VS, Giranda V, Chen X, Laver WG, Kohlbrenner W, Norbeck DW (2001) Novel {alpha}- and {beta}-amino acid inhibitors of influenza virus neuraminidase. Antimicrob Agents Chemother 45:2563–2570
99. DeGoey D, Chen H-J, Flosi WJ, Grampovnik DJ, Yeoung CM, Klein LL, Kempf DJ (2002) Enantioselective synthesis of antiinfluenza compound A-315675. J Org Chem 67:5445–5453
100. Barnes DM, Bhagavatula L, DeMattei J, Gupta A, Hill DR, Manna S, McLaughlin MA, Nichols P, Premchandran R, Rasmussen MW et al (2003) A highly diastereoselective vinylogous Mannich condensation and 1,4-conjugate addition of (Z)-propenyl cuprate in the synthesis of an influenza neuraminidase inhibitor. Tetrahedron Asymmetry 14: 3541–3551
101. Hanessian S, Bayrakdarian M, Luo X (2002) Total synthesis of A-315675:â€‰ a potent inhibitor of influenza neuraminidase. J Am Chem Soc 124:4716–4721
102. Ryu YB, Kim JH, Park S-J, Chang JS, Rho M-C, Bae K-H, Park KH, Lee WS (2010) Inhibition of neuraminidase activity by polyphenol compounds isolated from the roots of Glycyrrhiza uralensis. Bioorg Med Chem Lett 20:971–974
103. Jeong HJ, Ryu YB, Park S-J, Kim JH, Kwon H-J, Kim JH, Park KH, Rho M-C, Lee WS (2009) Neuraminidase inhibitory activities of flavonols isolated from Rhodiola rosea roots and their in vitro anti-influenza viral activities. Bioorg Med Chem 17:6816–6823
104. Ryu YB, Curtis-Long MJ, Lee JW, Ryu HW, Kim JY, Lee WS, Park KH (2009) Structural characteristics of flavanones and flavones from Cudrania tricuspidata for neuraminidase inhibition. Bioorg Med Chem Lett 19:4912–4915
105. Mercader AG, Pomilio AB (2010) QSAR study of flavonoids and biflavonoids as influenza H1N1 virus neuraminidase inhibitors. Eur J Med Chem 45:1724–1730
106. Hung H-C, Tseng C-P, Yang J-M, Ju Y-W, Tseng S-N, Chen Y-F, Chao Y-S, Hsieh H-P, Shih S-R, Hsu JTA (2009) Aurintricarboxylic acid inhibits influenza virus neuraminidase. Antiviral Res 81:123–131
107. Grienke U, Schmidtke M, Kirchmair J, Pfarr K, Wutzler P, Duìrrwald R, Wolber G, Liedl KR, Stuppner H, Rollinger JM (2009) Antiviral potential and molecular insight into neuraminidase inhibiting diarylheptanoids from alpinia katsumadai. J Med Chem 53: 778–786
108. An J, Lee DCW, Law AHY, Yang CLH, Poon LLM, Lau ASY, Jones SJM (2009) A novel small-molecule inhibitor of the avian influenza H5N1 virus determined through computational screening against the neuraminidase. J Med Chem 52:2667–2672
109. Shibazaki M, Tanaka K, Nagai K, Watanabe M, Fujita S, Suzuki K, Okada G, Yamamoto T (2004) YM-92447 (spinosulfate A), a neuraminidase inhibitor produced by an unidentified pycnidial fungus. J Antibiot 57:812–815
110. Collins P, Haire LF, Liu J, Russel RJ, Walker PA, Skehel JJ, Martin SR, Hay AJ, Gamblin SJ (2008) Crystal structures of oseltamivir-resistant influenza virus neuraminidase mutants. Nature 453:1258–1262
111. Amaro RE, Minh DDL, Cheng LS, Lindstrom WM, Olson AJ, Lin J-H, Li WW, McCammon JA (2007) Remarkable loop flexibility in avian influenza n1 and its implications for antiviral drug design. J Am Chem Soc 129:7764–7765
112. Mitrasinovic P (2009) On the Structure-Based design of novel inhibitors of H5N1 influenza A virus neuraminidase (NA). Biophys Chem 140:35–38
113. Rungrotmongkol T, Frecer V, De-Eknamkul W, Hannongbua S, Miertus S (2009) Design of oseltamivir analogs inhibiting neuraminidase of avian influenza virus H5N1. Antiviral Res 82:51–58
114. Park JW, Jo WH (2009) Computational design of novel, high-affinity neuraminidase inhibitors for H5N1 avian influenza virus. Eur J Med Chem 45:536–541

Clinical Experience with Influenza Virus Sialidase Inhibitors

Philippe Buchy

Abstract The clinical experience of influenza virus sialidase inhibitors has been relatively limited, although over the past 5 years a number of studies have been undertaken that demonstrate the efficacy and indeed the potential issues of these drugs. This chapter provides an overview of aspects of this clinical experience including the current clinical use of these drugs, role of laboratory, pharmacological aspects of these antiviral drugs, as well as treatment and chemoprophylaxis in different infection scenarios (e.g. seasonal epidemics versus pandemics).

1 Introduction

Influenza A (H3N2) and (H1N1) viruses are recurrent respiratory pathogens causing 250,000–500,000 deaths annually worldwide [1]. Rates of influenza-associated mortality are higher for elderly patients than for patients in any other age group, and rates of influenza-associated hospitalization are highest for children <5 years of age and adults >65 years of age [2, 3]. Recent events, including human cases of H5N1 avian influenza and pandemic A (H1N1) 2009 influenza, have intensified awareness of the threat of pandemic and triggered the development of plans for epidemic and pandemic control.

Vaccination is the primary strategy for the prevention of influenza, but there are a number of likely scenarios for which vaccination is inadequate and effective antiviral drugs would be of great importance. In particular, during the course of a pandemic, vaccine supplies may not be immediately available. In addition, during any influenza season, vaccine failures occur and sometimes antigenic drift in the virus may occur after the formulation of the new vaccine, consequently making it

P. Buchy (✉)
Institut Pasteur in Cambodia, Virology Unit, 5 Monivong blvd, PO Box 983, Phnom Penh, Cambodia
e-mail: pbuchy@pasteur-kh.org

M. von Itzstein (ed.), *Influenza Virus Sialidase - A Drug Discovery Target*,
Milestones in Drug Therapy, DOI 10.1007/978-3-7643-8927-7_7,

less protective. Anti-influenza drugs thus form an important part of a rational approach to influenza epidemics and are critical to responding to a pandemic or infections by novel strains.

Neuraminidase (or sialidase) inhibitors with activity against influenza viruses, specifically oseltamivir or zanamivir, are useful adjuncts in the prevention of influenza and effective when used early in the course of illness for treatment. Zanamivir (Relenza®, GlaxoSmithKline) and oseltamivir (Tamiflu®, Roche) are active against both influenza A and B viruses. Indeed, as a class, neuraminidase inhibitors are effective against all influenza A neuraminidase subtypes and, therefore, against all subtypes of influenza that might pose a pandemic threat. This represents a significant advantage over anti-influenza drugs like adamantanes (M2 ion channel inhibitors) and allows their use in both epidemic and pandemic situations.

The two medications differ in pharmacokinetics, adverse events, routes of administration, approved age groups, dosages, and costs.

2 Indications of Neuraminidase Inhibitors

Oseltamivir is licensed for treating influenza in patients aged $\geq$1 year, and zanamivir is licensed for the treatment of patients aged >7 years (>5 years in Japan) [4, 5]. The safety of oseltamivir in infants younger than 1 year old has not been rigorously proven. A concern had been raised after the observation that juvenile rats accumulated extremely high levels of oseltamivir in the central nervous system [6]. The immature blood–brain barrier of the infant is permeable and might similarly allow access of the drug to the central nervous system. But recent studies have found no evidence for serious adverse events [7, 8]. Oseltamivir is now advised for treating infants less than 1 year old and neonates with pandemic H1N1 illness [9]. For chemoprophylaxis of influenza, oseltamivir is licensed for use in person aged $\geq$1 year, and inhaled zanamivir can be used in individuals aged $\geq$5 years.

In response to the 2009 pandemic, the Food and Drug Administration (FDA) has authorized the use of the unapproved peramivir intravenous (IV) to treat certain adult and paediatric patients with suspect or laboratory-confirmed 2009 H1N1 infection or infection due to non-subtypable influenza A virus suspected to be 2009 H1N1 based on community epidemiology [10]. To date, peramivir IV has not been allowed for the treatment of seasonal influenza A or B virus infections, for outpatients with acute uncomplicated 2009 H1N1 virus infection, or for pre- or post-exposure chemoprophylaxis of influenza [10].

3 Role of Laboratory

Virus isolates are needed to monitor antiviral resistance and the emergence of novel human influenza A virus subtypes that might pose a pandemic threat. For instance, the H275Y mutation on the N1 neuraminidase protein causes approximately a

300-fold or greater reduction in susceptibility to oseltamivir compared with a wild-type virus. Seasonal H1N1 viruses with this mutation spread globally during 2008–2009, and some sporadic isolates and, less often, clusters of pandemic 2009 H1N1 viruses with this mutation have been detected [11, 12]. This mutation also confers an 80-fold or greater reduction in susceptibility to peramivir but has no effect on zanamivir susceptibility [13]. The mutation Q136K in seasonal H1N1 virus has no effect on oseltamivir susceptibility but causes approximately a 300-fold and a 70-fold reduction in zanamivir and peramivir susceptibility, respectively [14]. However, this mutation has not been detected directly in clinical samples to date.

Development of resistance to neuraminidase inhibitors while on therapy occurs less often than resistance to adamantanes [15]. In a multicentre study in the United States, only 5% of children who received oseltamivir therapy developed in vitro resistance in influenza isolates cultured during therapy [16]. But in Japan, a study documented resistance of 18% in isolates cultured from 50 oseltamivir-treated children [17, 18].

WHO recommends a treatment with zanamivir of the patients who have severe or progressive clinical illness due to oseltamivir-resistant and M2 inhibitor-resistant virus [4].

Peramivir IV should not be used for treatment of 2009 H1N1 virus infection in patients with documented or highly suspect oseltamivir resistance and should be used with caution in patients with documented or highly suspect zanamivir resistance [10].

4 Pharmacology

4.1 Route of Administration and Pharmacokinetics

The route of administration differs between the available drugs. Oseltamivir is administered orally in capsule (30, 45 or 75 mg) or in oral suspension form which when constituted with water as directed contains 12 mg/mL of oseltamivir. Approximately 80% of orally-administered oseltamivir is absorbed systemically [19] and then metabolized, primarily by hepatic esterase, to oseltamivir carboxylate which is the active neuraminidase inhibitor. Oseltamivir carboxylate is widely distributed in the body and achieves relatively high plasma levels and thus can theoretically act outside the respiratory tract. The metabolite's half-life is 6–10 h and is excreted in the urine [20, 21].

Zanamivir is not bioavailable orally and is marketed as a dry powder that is self-administered via oral inhalation by using a plastic device (Diskhaler, GlaxoSmithKline) included in the package with the medicine. The intravenous form has recently been used and should be widely available soon. Between 7 and 21% of the orally inhaled zanamivir reaches the lungs and approximately 4–17% of the total amount is absorbed systemically [22, 23]. The concentration of the drug in respiratory tract secretions has been estimated to be more than 1,000 times as high as the 50% inhibitory concentration (IC_{50}) for neuraminidase. In addition, the inhibitory effect starts within 10 s and these two features are favourable in terms of reducing the likelihood of emergence of drug-resistant variant strains. The systemically absorbed zanamivir has a half-life of 2.5–5.1 h and is excreted unchanged in the urine [22, 24]. Nevertheless, in a highly compromised patient with a focal pulmonary infiltrate, authors suggest that a limited distribution of the drug in the peripheral lungs might have compromised the complete viral clearance observed [25].

The intravenous form of zanamivir was used in September 2009 in a 10-year-old girl with acute lymphoblastic leukaemia who became critically ill after infection with a 2009 pandemic influenza A (H1N1) virus resistant to oseltamivir. The patient received 600 mg of zanamivir every 12 h intravenously for 15 days and successfully recovered without any zanamivir-related adverse effect [26].

Peramivir must be administered intravenously. The half-life of peramivir following administration of 0.5–8 mg/kg as a single dose of 4 mg/kg twice daily for 1 day ranged from 7.7 to 20.8 h. The major route of elimination of peramivir is via kidney (renal clearance accounts for ~90% of total clearance). The median IC_{50} values are 0.2 nM (range 0.09–1.4 nM) for influenza A strains, 1.3 nM (range 0.60–11 nM) for influenza B strains, from 0.06 to 0.26 nM for 2009 pandemic H1N1 viruses [10].

4.2 Adverse Events

Nausea and vomiting or abdominal pain were reported in approximately 9–10% of adult patients receiving oseltamivir treatment or prophylaxis versus 3–6% among persons receiving placebo [20, 27–29]. These adverse effects might be less severe when oseltamivir is taken with food and this does not interfere with the absorption of the drug [20]. Most adverse events occur only once close to the initiation of therapy and resolve spontaneously within 1 or 2 days [29]. Oseltamivir is generally quiet well tolerated even at single or divided daily dosages of $\leq$1,000 mg per day, although gastrointestinal symptoms seemed to occur at the highest doses [17, 30]. Transient neuropsychiatric events such as self-injury, hallucinations, abnormal behaviour or delirium have been reported in the setting of encephalitis or encephalopathy but also without obvious severe disease, especially among adolescents and adults living in Japan, and more frequently in males than females [31, 32]. Accurate data on the incidence of these events are not available but they seem to be in a range of 1 in 10,000–100,000 treatment courses. Two large U.S. claim databases indicate that the risk of neuropsychiatric adverse events in oseltamivir-treated patients is no higher than in patients receiving no antiviral treatment. This evidence strongly supports that the events are more likely to be caused by the disease itself rather than by oseltamivir [32].

Because of limited data available, zanamivir is not recommended for persons with underlying respiratory or cardiac disease [33]. Bronchospasm was observed after administration of zanamivir in patients with mild or moderate asthma [22]. Patients with lung disease should have a fast-acting inhaled bronchodilator available which should be used prior to administration of zanamivir [33]. Zanamivir powder should not be reconstituted in any liquid formulation and is not recommended for use in any nebulizer or mechanical ventilator because of the risk of obstruction of ventilator equipment [34]. During post-marketing surveillance, cases of respiratory function deterioration, cough and allergic reactions have been reported after inhalation of zanamivir [22, 35, 36]. Other adverse events were reported by clinical studies but the frequencies of these symptoms were similar for persons receiving placebo [37–39]. There have been also post-marketing reports, mostly from Japan, of delirium and abnormal behaviour leading to injury in patients with influenza who were receiving zanamivir. These events were reported primarily among children and often had an abrupt onset and rapid resolution but the contribution of zanamivir to these events has not been established. Nevertheless, patients with influenza should be closely monitored for signs of abnormal behaviour.

A total of 478 patients were included in clinical trials for peramivir IV and received only a single dose [10]. Gastrointestinal adverse events (diarrhoea, nausea, vomiting) were reported in 33% of patients receiving peramivir IV 200 mg, 28% of patients receiving peramivir IV 400 mg, and 15% of patients receiving oseltamivir [10]. No serious allergic-like reactions have been reported in clinical trials to date. Because peramivir IV is also a neuraminidase inhibitor, neuropsychiatric adverse events could always occur and patients should be closely monitored.

4.3 Neuraminidase Inhibitors and Pregnancy

No clinical studies have been conducted to assess the safety of oseltamivir and zanamivir during pregnancy but no specific problems were identified in pre-clinical testing. Thus, these drugs should only be used during pregnancy if the potential benefit justifies the potential risk to the embryo or foetus [20, 22].

Reports of malformation (1.1%) in a population of 90 pregnant Japanese women who received oseltamivir were within the incidence of major malformations in the general population. Tanaka et al. described the outcomes of four pregnant women who were exposed to zanamivir, with one pregnancy spontaneously miscarried, one terminated and two healthy children delivered [40].

4.4 Drug Interactions

Clinical data are limited regarding drug interactions with oseltamivir and zanamivir. The concurrent use of oseltamivir with live attenuated influenza vaccine delivered intranasally has not been evaluated. However, because of the potential for interactions between these products, live attenuated influenza vaccine should not be administered within 2 weeks before or 48 h after administration of oseltamivir. Inactivated influenza vaccine can be administered at any time relative to the use of oseltamivir. Because oseltamivir and its active metabolite are excreted in the urine by glomerular filtration and tubular secretion, interactions with other drugs excreted by this pathway could potentially exist. Thus, co-administration of oseltamivir and probenicid results in reduced clearance of oseltamivir carboxylate by approximately 50% and in an approximate twofold increase in the plasma levels of oseltamivir carboxylate [41]. The effect of probenicid has been suggested as a means of dose-sparing. However, one pharmacokinetic study found that probenicid might reduce penetration of oseltamivir carboxylate into respiratory secretions [42].

For zanamivir, no known drug interactions have been reported, and no critical drug interactions have been predicted on the basis of in vitro and animal studies [22, 43]. Indeed, zanamivir is not a substrate nor does it affect cytochrome P450 isoenzymes in human liver microsomes. As for oseltamivir, because of the potential for interactions between live attenuated influenza vaccine and the neuraminidase inhibitor, such vaccine should not be administered within 2 weeks before or 48 h after administration of zanamivir. Trivalent inactivated influenza vaccine can be administered at any time relative to the use of zanamivir.

Peramivir IV is primarily eliminated by kidneys. Co-administration of peramivir IV with drugs that reduce renal function or compete for active tubular secretion may increase plasma concentration of peramivir and/or increase the concentrations of other renally eliminated drugs [10].

5 Treatment and Chemoprophylaxis

5.1 Pandemic 2009 A (H1N1) Virus

5.1.1 Treatment

In April 2009, a new influenza A (H1N1) with a swine origin appeared in Mexico and in the United States. The rapid spread of the virus led WHO to declare on 11 June 2009 that the first influenza pandemic in 41 years had occurred.

Vaccines that are effective for a newly arising strain require several months of development. Thus, antiviral drugs form an important part of a strategy during a pandemic. The protective efficacy of the neuraminidase inhibitors during a pandemic would be expected to be at least as high as that of the adamantanes that were found to have a protective efficacy of around 70% during the 1968 and 1977 pandemics [44, 45]. The pandemic H1N1 virus is resistant to adamantanes, so these drugs have no role in management at present.

These guidelines have to be placed in the context of epidemiological and antiviral susceptibility data on the locally circulating influenza strains [4]. WHO recommends to treat patients who have severe or progressive clinical illness with oseltamivir. Treatment should be initiated as soon as possible (laboratory confirmation is not necessary for the initiation of the treatment) since in a recent study, patients who were admitted to an intensive care unit (ICU) or died were less likely to have received neuraminidase inhibitors within 48 h after the onset of symptoms. Experience from use in patients with H5N1 virus infection and severe lower respiratory tract disease suggests that later initiation of treatment may also be effective, whenever viral replication is present or strongly suspected [4].

So, despite the absence of definitive data regarding clinical effectiveness, treatment with antiviral drugs should be initiated in hospitalized patients with 2009 H1N1 infection even if such therapy is initiated more than 48 h after the onset of symptoms, especially in patients with pneumonia and outpatients who are at increased risk of complications, including pregnant women [46]. Consideration may be given to the use of higher doses up to 150 mg twice daily (in adults) and longer duration of treatment (at least 10 days) depending on the clinical response (i.e. severe or progressive illness not responding to normal treatment regimen). Since the virus can cause severe illness and death in pregnant and post-partum women [47], pregnant women should be treated, although there is not sufficient safety data for doses higher than 75 mg twice daily [48].

In situations where oseltamivir is not available or not possible to use, or if the virus is resistant to oseltamivir, patients who have severe or progressive clinical illness could be treated with zanamivir. This drug, in its inhalation dosage form, is difficult to administer to severely ill patients. If available, an intravenous formulation of zanamivir would be the preferred form in such patients and has been used in salvaging immunocompromised hosts infected with oseltamivir-resistant virus [26].

WHO recommends to treat all patients with uncomplicated illness in "at-risk" groups (infants and children aged less than 5, elderly >65 years, nursing home residents, pregnant women, patients with chronic co-morbid conditions such as cardiovascular, respiratory or liver disease, diabetes, and those with immunosuppression related to malignancy, HIV infection or other diseases) with oseltamivir or zanamivir. Patients not in "at-risk" groups who have uncomplicated illness due to a confirmed or strongly suspected influenza virus infection need not be treated with antivirals. Standard doses of these antiviral drugs are provided in Table 1.

For infants less than 1 year of age, WHO recommends to treat the patients with doses as described in Table 2 [48] (WHO Clinical management of human infection with pandemic H1N1 2009, revised guidance)

As of October 2009, the U.S. Food and Drug Administration issued an Emergency Use Authorization (EUA) for treatment with peramivir IV of hospitalized patients with 2009 H1N1 influenza who have potentially life-threatening suspected or laboratory-confirmed infection [49].

One hundred per cent of the pandemic (H1N1) 2009 influenza virus are resistant to M2 inhibitors and since July 2009 oseltamivir-resistant strains are regularly detected [10], hence the importance to develop and licence new anti-influenza

Table 1 Standard dose of oseltamivir and zanamivir recommended for treatment of pandemic H1N1-infected patients

Age groups (years)						
Duration	1–4	5–6	7–9	10–12	13–64	>65
Oseltamivir						
5 days	Weight-adjusted doses: – 30 mg twice daily for $\leq$ 15 kg – 45 mg twice daily for >15–23 kg – 60 mg twice daily for >23–40 kg – 75 mg twice day for >40 kg				75 mg twice daily	75 mg twice daily
Zanamivir						
5 days	Not licensed for use	10 mg (2 inhalations) twice daily			10 mg (2 inhalations) twice daily	10 mg (2 inhalations) twice daily

Adapted from [4]

Table 2 Standard dose of oseltamivir recommended for treatment of pandemic H1N1-infected infants

>3–12 months	3 mg/kg twice daily
>1–3 months	2.5 mg/kg twice daily
0–1 month[a]	2 mg/kg twice daily

[a]There are no data available regarding the administration of oseltamivir in infants less than 1 month of age
Adapted from [48]

drugs such as favipiravir (T705) and laninamivir (CS-8958) which will increase the potential treatment options [50, 51].

5.1.2 Chemoprophylaxis

Antiviral chemoprophylaxis of pandemic H1N1 influenza should generally be considered with regard to the benefit of providing short-term protection from illness and possibly infection, and the economic impact. Chemoprophylaxis is generally not recommended, as the opportunity cost and utilization of antiviral drugs that may be needed for treatment is not warranted. The threshold number of individuals who would need to be treated (number needed to treat or NNT) to prevent a single case of influenza to balance the trade-off between use of resources and benefits was evaluated. It was determined that an NNT of 20 or less may be reasonable in certain situations associated with more serious consequences from infection [4]. The use of chemoprophylaxis assumes that other control measures (i.e. infection control) are in place, vaccination is planned, mechanisms for delivery of drugs and costs are acceptable. Antiviral chemoprophylaxis may have particular benefits in the higher-risk situation settings. For example, settings where a high proportion of the community falls within one or more at-risk groups such as residential healthcare institutions: nursing homes, certain hospital wards and hospices; discrete closed or semi-closed settings with a high level of person-to-person contact and finite healthcare resources, where a high attack rate over a short time span may overload healthcare resources resulting in high mortality and morbidity; combination of the above two settings such as refugee camps or disaster zones. Chemoprophylaxis may also have benefits in high gain groups: groups with high exposure to infection but whose function is crucial to mitigation of the epidemic, and where a high attack rate over a short period of time would severely compromise health service provision (i.e. healthcare workers); other groups that are critical to maintaining core functions, with a high level of person-to-person contact, where a high attack rate other a short time span would likely compromise service function. Antiviral chemoprophylaxis may also be recommended to high-risk individuals: individuals whose circumstances lead to a particular high risk of complication from influenza, and with a high risk of exposure (i.e. stem cell transplantation or solid organ transplant at a time of high local prevalence of infection) [4].

The aim of the chemoprophylaxis is to reduce the probability of infection at a time of exceptional vulnerability, and particularly to be able to provide protective cover until vaccination can be introduced or can become effective. WHO recommends the use of antivirals as chemoprophylaxis of the pandemic (H1N1) 2009 influenza where the risk of human-to-human transmission of influenza is high or low and the likelihood of complications of infection is high (either due to the strain or baseline risk of the exposed group). Oseltamivir or zanamivir might then be used as post-exposure chemoprophylaxis for the affected community or group, individuals in "at-risk" groups or healthcare workers. If the likelihood of complications of infection is low, antiviral chemoprophylaxis need not be offered to individuals in "at-risk" groups or healthcare workers. This recommendation applies independent of risk of human-to-human transmission [4]. Recommended doses for chemoprophylaxix are indicated in Tables 3 and 4.

Table 3 Dosage recommendations for chemoprophylaxis in pandemic H1N1-infected patients

Age groups (years)						
Duration	1–4	5–6	7–9	10–12	13–64	>65
Oseltamivir						
10 days	Weight-adjusted doses: – 30 mg/day for ≤15 kg – 45 mg/day for >15–23 kg – 60 mg/day for >23–40 kg – 75 mg/day for >40 kg				75 mg/day	75 mg/day
Zanamivir						
10 days	Not recommended	10 mg (2 inhalations once daily)	10 mg (2 inhalations once daily)	10 mg (2 inhalations once daily)	10 mg (2 inhalations once daily)	10 mg (2 inhalations once daily)

Adapted from [4, 49]

Table 4 Dosage recommendation for chemoprophylaxis in pandemic H1N1-infected infants

>3–12 months	3 mg/kg/day for 10 days
<3 months	Not recommended unless situation judged critical due to limited data on use in this age group

Adapted from [4, 49]

5.2 *Avian Influenza A (H5N1) Virus*

5.2.1 Treatment

The highly pathogenic avian influenza A/H5N1 virus is not easily transmitted from human to human but an adaptation to human host could always happen. The H5N1 viruses are often resistant to the adamantanes (especially all the clade 1 viruses) but are usually sensitive to neuraminidase inhibitors. Thus, many countries have reportedly stockpiled oseltamivir as part of their pandemic planning. Emergence of oseltamivir-resistant strains or observation of reduction in oseltamivir efficacy has already been reported [52]. This observation suggests that in addition to oseltamivir, zanamivir should be included as part of pandemic preparedness.

The guidelines for the pharmacological management of humans infected with the highly pathogenic avian influenza A (H5N1) virus provided by WHO are similar to those recommended for treating patients infected with the pandemic 2009 A (H1N1) virus [4]. However, the H5N1 virus is not always resistant to adamantanes, so that these drugs may have a potential role in avian influenza infection management.

Early treatment with oseltamivir is associated with improved survival [52]. The clade 1 viruses isolated since 2004 generally show enhanced susceptibility to oseltamivir by comparison to strains detected in Hong Kong in 1997 [53]. But the high level of replication of some oseltamivir-susceptible strains requires higher doses or more prolonged administration, or both, in animal models [54, 55]. Clade 2 viruses isolated from Indonesia and Turkey appear to be 15–30 times less sensitive to oseltamivir than clade 1 viruses but the relevance of such differences in oseltamivir susceptibility remains to be determined [55, 56]. During oseltamivir therapy, the emergence of resistant variants with a mutation H274Y on the neuraminidase protein was associated with a fatal outcome in two patients [57]. The mutation N294S on the neuraminidase is associated with a reduction in oseltamivir susceptibility by a factor of 12–15 times and was reported in two fatal Egyptian cases before receipt of oseltamivir [58]. As for pandemic H1N1 influenza, early treatment with oseltamivir is recommended and some data suggest that it improves survival [52, 59, 60]. A higher dose of oseltamivir (e.g. 150 mg twice daily in adults) and an increased duration for therapy for a total of 10 days is often applied (Table 1), given the high levels of replication of the H5N1 virus, observations of progressive disease despite early administration of standard doses (75 mg twice daily for 5 days), and the proven safety of higher doses in adults with seasonal influenza [60]. In areas where influenza A (H5N1) viruses are likely to be susceptible to amantadine, combination treatment with oseltamivir and adamantanes would be reasonable, especially in serious cases [52]. Indeed, in mouse models, the combination of oseltamivir and amantadine significantly increased survival rates as compared with monotherapy [61].

Zanamivir is active against oseltamivir-resistant variants with H274Y or N294S mutations on the neuraminidase but the value of this inhaled drug has not been studied in H5N1-infected patients. The intravenous formulation of the zanamivir, successfully experimentally evaluated on macaques [62], will probably bring an important added value for the treatment of oseltamivir-resistant strains and the severe form of the disease. Peramivir efficiently reduced H5N1 virus replication and mitigated disease in mice and ferrets [63, 64]. In addition, in vitro studies suggest that a majority of H5N1 strains are fully sensitive to this new neuraminidase inhibitor [65].

5.2.2 Chemoprophylaxis

For chemoprophylaxis purposes, oseltamivir or zanamivir should be used for 7–10 days after the last exposure in high risk group, including pregnant women (household or close family contacts of a strongly suspected or confirmed H5N1 patient) and in moderate risk group, including pregnant women (personnel involved in handling sick animals or decontaminating affected environments if personal protective equipment may not have been used properly; individuals with unprotected or very close direct exposure to sick or dead animals infected with the H5N1 virus; healthcare personnel in close contact with strongly suspected or confirmed H5N1 patients, for example during intubation or performing tracheal suctioning, or delivering nebulized drugs, or handling inadequately screened/sealed body fluids with insufficient personal protective equipment; laboratory personnel who might have an unprotected exposure to virus-containing samples) [59].

5.3 *Seasonal Influenza Viruses A (H3N2), A (H1N1) and B*

5.3.1 Treatment

Antiviral treatment should ideally be initiated within 2 days of illness, although the benefit of treatment is greater as the time after illness is reduced. When zanamivir or oseltamivir is administered within 36–48 h after onset of uncomplicated influenza illness among otherwise healthy adults, a decrease in symptomatic illness of 1–2 days occurs [28, 29, 37–39, 66–71]. In an important trial, oseltamivir reduced the time to return to normal activity by 2.8 days [28]. These finding are important to the economic impact of influenza and, perhaps, to planning for pandemics. Oseltamivir treatment starting within 12 h after the onset of fever shortened the illness by more than 3 days, as compared with treatment that was started at 48 h [70]. In a review including six placebo-controlled trials of oseltamivir in "healthy" adults, six trials in "at-risk" patients (including patients with co-morbidities, elderly and children aged 6–12 years), two trials in children and three trials in elderly, oseltamivir was associated with a reduction of 16.3 h in the duration of symptoms in an intention-to-treat (ITT) population and of 22.7 h in an infected population. The time to resume normal activity was 34.8 h for the ITT population and 36.3 h for an infected population [72]. In the same review, the reported results demonstrated that zanamivir was associated with a reduction of 17 h (0.71 days) for an ITT population and 27.7 h (1.07 days) in an infected population. The time to resume normal activity showed no statistically significant advantage for zanamivir compared to placebo [72].

Minimal or no benefit was reported when antiviral treatment is initiated $\geq$2 days after onset of uncomplicated influenza. After neuraminidase inhibitors treatment, the duration of viral shedding was significantly reduced in children treated by oseltamivir [73]. In adults, viral titres decrease rapidly by day 3–4 of illness in untreated previously healthy individuals and this may explain why there is a limited window of efficacy for the initiation of treatment. Viral loads are higher and shedding more persistent in infants, young children, and immunocompromised hosts [74].

Data are limited about effectiveness of zanamivir and oseltamivir in preventing serious influenza-related complications (e.g. bacterial or viral pneumonia or exacerbation of chronic diseases) or for preventing influenza among persons at high risk of complications of influenza. A meta-analysis of trials involving influenza-positive high-risk patients older than 65 years of age or with chronic medical conditions concluded that zanamivir reduced the time of symptom attenuation by 2 days and that oseltamivir did so by about 0.5 day [66]. In a retrospective pooled analysis of data from high-risk patients, zanamivir-treated patients had a treatment benefit of 2.5 days compared to those with placebo and returned to normal activities 3 days earlier [75]. In a study including children and adolescents who were considered to be at risk of influenza complications because they had an active chronic medical condition within the 12 months preceding influenza diagnosis, patients treated by oseltamivir had, in the 14 days after influenza diagnosis, 26 and 31% reduction in the risks of respiratory illness other than pneumonia and otitis media, respectively [76]. Experiments in animals suggest that neuraminidase inhibitors might reduce the incidence of secondary bacterial pneumonia by interfering with the synergism between influenza neuraminidase and *Streptococcus pneumoniae* [77].

In a study combining 10 clinical trials, the risk of pneumonia among those participants with laboratory-confirmed influenza receiving oseltamivir was approximately 50% lower than among those persons receiving a placebo and 34% lower among patients at high risk for complications [78]. In children from 1 to 12 years of age treated with oseltamivir no more than 48 h after onset of the illness, the length of the illness was shortened by 36 h and there was a decreased incidence of otitis media by 44% [16]. In children with asthma, a greater improvement in lung function and fewer asthma exacerbations were observed among patients treated with oseltamivir compared with those who received a placebo [79].

Three studies suggest that oseltamivir reduces severe clinical outcome in patients hospitalized with seasonal influenza [80–82]. Oseltamivir was associated with a significantly decreased risk of death within 15 days of hospitalization in patients with an average age of 77 years. Benefit was observed even amongst those starting treatment >48 h after symptom onset [81]. A retrospective cohort study conducted in Hong Kong reported that persons who received oseltamivir treatment >48 h of illness onset had a median length of stay of 6 days compared with 4 days for persons who received oseltamivir within 48 h of symptom onset [82]. In a prospective study, the same authors showed that patients hospitalized with severe influenza had more active and prolonged viral replication, and that oseltamivir treatment that started within the first 4 days of illness enhanced viral clearance [82].

Human clinical studies indicate that zanamivir and oseltamivir have activities against influenza B viruses. Children treated with oseltamivir for influenza A virus infection resolve fever and stop shedding virus more quickly than children with influenza B, suggesting that oseltamivir might be less effective for the treatment of influenza B [83, 84]. In a double-blinded study, oseltamivir had lower clinical effectiveness against influenza A (H1N1) than against influenza A (H3N2) [84].

The U.S. Centers for Disease Control and Prevention and the Infectious Diseases Society of America have identified seasonal influenza patients (adults and children) with a high priority for treatment: (1) patients hospitalized with laboratory-confirmed influenza (with limited data suggesting that the treatment could be of benefit, even in persons whose antiviral treatment is initiated >48 h after illness onset), (2) patients with laboratory-confirmed influenza pneumonia, (3) patients with laboratory-confirmed influenza and bacterial coinfections, (4) persons with laboratory-confirmed influenza who are at high risk of influenza complications (little data are available by which to make recommendations regarding treatment of persons >48 h after symptom onset), (5) patients with laboratory-confirmed influenza within 48 h of illness onset who want to decrease the duration or severity of their symptoms and transmission of influenza to others at higher risk of complications (those whose onset of symptoms occurred >48 h before presentation with persisting moderate to severe illness may also benefit from treatment, but safety and efficacy in this population have not been evaluated adequately). However, treatment does not need to be limited to those persons [49, 85].

WHO, in its guidelines for treatment of patients with confirmed or strongly suspected infection with seasonal influenza virus, recommends that patients who have severe or progressive clinical illness due to an oseltamivir-susceptible and M2 inhibitor-susceptible virus might be treated with both oseltamivir and either amantadine or rimantadine. In situations where there is co-circulation influenza A virus subtypes (even if there is probable or known oseltamivir resistance), all patients (including pregnant women) who have severe or progressive clinical presentation should be treated with oseltamivir and either amantadine or rimantadine. If laboratory data confirm drug resistance in the infecting strain, then the inactive drug should be stopped [4].

Patients not in "at-risk" groups who have uncomplicated illness due to confirmed or strongly suspected influenza virus infection need not be treated with antivirals. But in the situation where there are co-circulating influenza A virus subtypes (even when these include probable or known oseltamivir resistance), patients in "at-risk" groups (except pregnant women) with uncomplicated illness due to confirmed or strongly suspected seasonal influenza virus infection should be treated with zanamivir, or with oseltamivir plus amantadine or rimantadine. For pregnant women with uncomplicated illness due to infection with seasonal influenza virus probably or known to be oseltamivir-resistant, zanamivir is the treatment of choice.

Table 5 Standard dosage recommendations for treatment of seasonal influenza

	Age groups (years)					
Duration	1–4	5–6	7–9	10–12	13–64	>65
Oseltamivir						
5 days	Weight-adjusted doses[1]: – 30 mg/day for ≤15 kg – 45 mg/day for >15–23 kg – 60 mg/day for >23–40 kg – 75 mg/day for >40 kg				75 mg twice daily[a]	75 mg twice daily[a]
Zanamivir						
5 days	Not licensed for use	10 mg (2 inhalations) twice daily			10 mg (2 inhalations) twice daily	10 mg (2 inhalations) twice daily

[a]Reduction in dose of oseltamivir is recommended for persons with creatinine clearance <30 mL/min
Adapted from [4]

Indeed, pregnant women and children aged less than 1 year with uncomplicated illness due to influenza infection should not be treated with M2 inhibitors [4].

Where the most prevalent virus is probably or known to be oseltamivir-resistant, immunosuppressed patients with seasonal influenza infection should be treated with zanamivir plus rimantadine [4].

In adults and children aged >5 years, the recommended dosage of zanamivir for treatment is 10 mg (2 inhalations, each blister containing 5 mg of active drug) twice daily for 5 days [4].

Oseltamivir is licensed for treatment of uncomplicated acute influenza A or B infections. In adults, the recommended dosage is 75 mg twice daily during 5 days. For patients with creatinine clearance of 10–30 mL per minute, reduction of the treatment dosage of oseltamivir to 75 mg once daily is recommended [4]. Standard dosage recommendations for treatment of seasonal influenza are indicated in Table 5.

5.3.2 Chemoprophylaxis

Oseltamivir and zanamivir have similar efficacy in preventing community-acquired influenza illness in healthy adults: 82 and 84%, respectively [27, 86]. These neuraminidase inhibitors also prevent influenza illness after a household member had influenza diagnosed (efficacy: zanamivir, 72–82%; oseltamivir, 68–89%) [27, 86–89]. A moderate to excellent efficacy was observed for the prevention of influenza in institutional settings [90–92]. In a double-blind randomized study, the use of oseltamivir for seasonal prophylaxis in residential homes for elderly persons led to a 92% reduction in the incidence of laboratory-confirmed influenza, even though the great majority of the elderly residents had received the appropriate vaccine for the season [92]. Thus, early recognition of influenza symptoms in the elderly and a rapid response by medical staff could significantly enhance the effectiveness of oseltamivir prophylaxis for outbreaks control in institutions [90, 93].

In children aged 1 year or older, the protective efficacy of oseltamivir administered after exposure could be as high as 55% when compared with treating-only index cases [87].

The efficacy of antiviral agents in preventing influenza among severely immunocompromised patients is poorly known. But oseltamivir appeared to be safe and well tolerated in managing influenza outbreak in a haematopoietic stem cell transplantation outpatient residence [94].

When influenza vaccine is administered after vaccine activity and since development of antibodies in adults takes approximately 2 weeks [95, 96], chemoprophylaxis should be considered for persons at high risk for influenza complications.

To reduce the spread of virus to persons at high risk, chemoprophylaxis during peak influenza activity can be considered for unvaccinated persons (e.g. hospital's employees, clinical staff, household members, visiting nurses, etc.) who have frequent contact with persons at high risk. Chemoprophylaxis can be considered for persons at high risk who are more likely to have an inadequate antibody response to influenza vaccine.

According to the U.S. Centers for Disease Control and Prevention and the Infectious Diseases Society of America, some individuals (adults and children $\geq$1 year) are at higher priority for chemoprophylaxis: (1) persons at high risk during the 2 weeks after influenza vaccination (after the second dose for children aged <9 years who have not been previously vaccinated) if influenza viruses are circulating in the community; (2) persons at high risk for complications in whom influenza vaccine is contraindicated; (3) family members or healthcare providers who are unvaccinated and are likely to have ongoing , close exposure to persons at high risk or unvaccinated persons or infants aged <6 months; (4) persons at high risk and their family members and close contacts, and healthcare workers when circulating strains of influenza virus in the community are not matched with vaccine strains; (5) persons with immune deficiencies or those who might not respond to vaccination (e.g. HIV-infected patients, persons receiving immunosuppressive medications), (6) unvaccinated staff and persons during response to an outbreak in a closed institutional setting with residents at high risk (e.g. extended-care facilities). However, chemoprophylaxis does not need to be limited to these individuals [49, 85].

Influenza viruses and their susceptibility to available antiviral drugs evolve rapidly. Clinicians should maintain familiarity with local patterns of influenza circulation in their community throughout the influenza season (i.e. updated information are available on WHO and CDC websites).

Individuals who are candidates for receiving chemoprophylaxis as part of an outbreak known to be caused by oseltamivir-resistant viruses can receive zanamivir. However, zanamivir is not licensed for prophylaxis in children aged <5 years [22]. Zanamivir is also not recommended for use in persons with chronic cardiopulmonary conditions. In those cases, a combination of oseltamivir and either amantadine or preferably rimantadine (because of its more favourable adverse effect profile) can be considered because viruses that are resistant to oseltamivir (i.e. H1N1) are often susceptible to M2 protein inhibitors. Adamantanes should not

Table 6 Standard dosage recommendations for chemoprophylaxis of seasonal influenza

Age groups (years)						
Duration	1–4	5–6	7–9	10–12	13–64	>65
Oseltamivir						
Begin as soon as exposure identified and continue for 5–7 days after last known exposure[a]	Weight-adjusted doses[b]: – 30 mg/day for ≤15 kg – 45 mg/day for >15–23 kg – 60 mg/day for >23–40 kg – 75 mg/day for >40 kg				75 mg/day	75 mg/day
Zanamivir						
Begin as soon as exposure identified and continue for 5–7 days after last known exposure[a]	Not licensed for use	10 mg (2 inhalations) once daily	10 mg (2 inhalations) once daily	10 mg (2 inhalations) once daily	10 mg (2 inhalations) once daily	10 mg (2 inhalations) once daily

[a]For control of outbreaks in long-term care facilities and hospitals, CDC recommends chemoprophylaxis for a minimum of 2 weeks, and up to 1 week after the last known case was identified
[b]Reduction in dose of oseltamivir is recommended for persons with creatinine clearance <30 mL/min
Adapted from [4, 49]

be used for influenza A (H3N2) chemoprophylaxis. Only oseltamivir or zanamivir should be used for influenza B chemoprophylaxis [49, 85].

In a recent systematic review which included six trials, data suggest that in adults there were statistically significantly fewer cases of laboratory-confirmed symptomatic infections in patients receiving oseltamivir compared to placebo, with RR = 0.27. There were also significantly fewer cases of infection in elderly individuals (RR = 0.08) and in mixed households including adults and children post-exposure prophylaxis resulted in RR = 0.19 [97]. In the same review, nine zanamivir trials were included and the analyses demonstrated a statistically significant benefit for zanamivir compared to placebo in all populations except the elderly, with a protective efficacy ranging from 70% to just over 80% [97].

In adults and children aged ≥5 years, the recommended dosage of zanamivir for chemoprophylaxis is 10 mg once daily. For chemoprophylaxis in household settings, zanamivir should be administered for 10 days and during community outbreaks the duration of the treatment should be 28 days.

Oseltamivir is licensed for chemoprophylaxis of influenza A or B infection in adults and children aged ≥1 year. In adults, the recommended dosage is 75 mg per day during up to 6 weeks. In children, the dosage varies by the weight of the child (see Table 6). For patients with creatinine clearance of 10–30 mL per minute, reduction of the dosage to 75 mg every other day is recommended. The standard dosage recommendations are indicated in Table 6.

Acknowledgements We would like to thank Prof. Frederick G. Hayden for his very valuable comments and kind assistance.

References

1. World Health Organization (2003, posting date) Fact sheet 211 Influenza. Available at: http://www.who.int/mediacentre/factsheets/fs211/en/index.html. Accessed 22 Sep 2009
2. Thompson WW, Shay DK, Weintraub E (2003) Mortality associated with influenza an respiratory syncitial virus in the United States. JAMA 289:179–86
3. Terebuth P, Uyeki T, Kukuda K (2003) Impact of influenza on young children and the shaping of United States influenza vaccine policy. Pediatr Infect Dis J 22:S231–235
4. World Health Organization. WHO guidelines for pharmacological management of pandemic (H1N1) 2009 influenza and other influenza viruses. Available at: http://www.who.int/csr/resources/publications/swineflu/h1n1_guidelines_pharmaceutical_mngt.pdf. Accessed 5 Jan 2010
5. Centers for Disease Control and Prevention (2008) Prevention and control of influenza. Recommendations of the Advisory Committee on Immunization Practices (ACIP), 2008. MMWR 57:1–61
6. Health Sciences Authority. Product safety alert: new preclinical findings on oseltamivir. March, 2004. Available at: http://www.hsa.gov.sg/publish/hsaportal/en/health_products_regulation/safety_information/product_safety_alerts/safety_alerts_2004/oseltamivir.html. Accessed 31 Jan 2010
7. Okamoto S, Kamiya I, Kishida K, Shimakawa T, Fukui T, Morimoto T (2005) Experience with oseltamivir for infants younger than 1 year old in Japan. Pediatr Infect Dis J 24:575–576
8. Toovey S, Rayner C, Prinssen E, Chu T, Donner B, Thakrar B, Dutkowski R, Hoffmann G, Breidenbach A, Lindermann L et al (2008) Assessment of neuropsychiatric adverse events in influenza patients treated with oseltamivir: a comprehensive review. Drug Saf 31:1097–1114
9. Fritsch BF. FDA update on the H1N1 flu vaccine and antiviral medication. Available at: http://www.fda.gov/downloads/ForHealthProfessionals/UCM185724.pdf. Accessed 1 Feb 2010
10. Centers for Disease Control and Prevention. Emergency use authorization of peramivir IV fact sheet for health care providers. Available at: http://www.cdc.gov/h1n1flu/eua/Final%20HCP%20Fact%20sheet%20Peramivir%20IV_CDC.pdf. Accessed 31 Jan 2010
11. World Health Organization (2009) Oseltamivir-resistant pandemic (H1N1) 2009 influenza virus, October 2009. Wkly Epidemiol Rec 84:453–468
12. Mai LQ, Wertheim HFL, Duong TN, van Doorn RH, Hien NT, Horby P, for the Vietnam H1N1 Investigation Team (2010) A community cluster of oseltamivir-resistant cases of 2009 H1N1 influenza. N Engl J Med 362:86–87
13. Hurt AC, Ernest J, Deng Y (2009) Emergence and spread of oseltamivir-resistant A(H1N1) influenza viruses in Oceania, South-East Asia and South Africa. Antiviral Res 83:90–93
14. Hurt AC, Holien JK, Parker M, Kelso A, Barr IG (2009) Zanamivir-resistant influenza viruses with a novel neuraminidase mutation. J Virol 83:10366–10373
15. Hayden FG (2006) Antiviral resistance in influenza viruses: implications for management and pandemic response. N Engl J Med 354:785–788
16. Whitley RJ, Hayden FG, Reisinger KS (2001) Oral oseltamivir treatment of influenza in children. Pediatr Infect Dis J 20:127–133
17. Moscona A (2004) Oseltamivir-resistant influenza? Lancet 364:733–734
18. Kiso M, Mitamura K, Sakai-Tagawa Y (2004) Resistant influenza A Viruses in children treated with oseltamivir: descriptive study. Lancet 364:759–765
19. Bardsley-Elliot A, Noble S (1999) Oseltamivir. Drugs 58:851–62
20. Roche Laboratories, Inc. (2005) Tamiflu (oseltamivir for phosphate) capsules and oral suspension [Package insert]. Roche Laboratories
21. He G, Massarella J, Ward P (1999) Clinical pharmacokinetics of the prodrug oseltamivir and its active metabolite Ro 64–0802. Clin Pharmacokinet 37:471–484
22. Glaxo Wellcome, Inc. (2001) Relenza (zanamivir for inhalation) [Package insert]. Research Triangle Park, NC: Glaxo Wellcome

23. Cass LM, Brown J, Pickford M (1999) Pharmacoscintigraphic evaluation of lung deposition of inhaled zanamivir in healthy volunteers. Clin Pharmacokinet 36:21–31
24. LaForce C, Man CY, Henderson FW (2007) Efficacy and safety of inhaled zanamivir in the prevention of influenza in community-dwelling, high-risk adult and adolescent subject: a 28-day, multicenter randomized, double-blind, placebo-controlled trial. Clin Ther 29:1579–1590
25. Medeiros R, Rameix-Welti MA, Lorin V, Ribaud P, Manuguerra JC, Socie G, Scieux C, Naffakh N, van der Werf S (2007) Failure of zanamivir therapy for pneumonia in a bone-marrow transplant recipient infected by a zanamivir-sensitive influenza A (H1N1) virus. Antivir Ther 12:571–576
26. Gaur AH, Bagga B, Barman S, Hayden R, Lamptey A, Hoffman JM, Bhojwani D, Flynn PM, Tuomanen E, Webby R et al (2010) Intravenous zanamivir for oseltamivir-resistant 2009 H1N1 influenza. N Engl J Med 362:88–89
27. Hayden FG, Treanor JJ, Friz RS (1999) Use of the oral neuraminidase inhibitor oseltamivir in experimental human influenza: randomized controlled trials for prevention and treatment. JAMA 282:1240–1246
28. Treanor JJ, Hayden FG, Vrooman PS (2000) Efficacy and safety of the oral neuraminidase inhibitor oseltaminir in treating acute influenza: a randomized controlled trial. US Oral Neuraminidase Study Group. JAMA 283:1016–1024
29. Nicholson KG, Aoki FY, Osterhaus AD (2000) Efficacy and safety of oseltamivir in treatment of acute influenza: a randomized controlled trial. Neuraminidase Inhibitor Flu Treatment Investigator Group. Lancet 355:1845–1850
30. Jefferson T, Demicheli V, Rivetti D, Jones M, Di Pietrantonj C, Rivetti A (2006) Antivirals for influenza in healthy adults: systematic review. Lancet 367:303–313
31. Anonymous (2007) New concerns about oseltamivir [Editorial]. Lancet 369:1056
32. Toovey S, Rayner C, Prinssen E, Chu T, Donner B, Thakrar B, Dutkowski R, Hoffmann G, Breidenbach A, Lindermann L et al (2008) Assessment of neuropsychiatric adverse events in influenza patients treated with oseltamivir: a comprehensive review. Drug Saf 31:1097–1114
33. Centers for Disease Control and Prevention. Relenza EUA, Fact sheet for Health Care Providers. Auhorized by FDA on October 30, 2009. Available at: http://www.cdc.gov/h1n1flu/eua/pdf/relenza_hcp_103009_fda_authorized_10.30.09.pdf. Accessed 1 Feb 2010
34. U.S. Food and Drug Administration. Relenza (zanamivir) inhalation powder. Available at: http://www.fda.gov/safety/medwatch/safetyinformation/safetyalertsforhumanmedicalproducts/ucm186081.htm. Accessed 31 Jan 2010
35. Gravenstein S, Johnston SL, Loeschel E (2001) Zanamivir: a review of clinical safety in individuals at high risk of developing influenza-related complications. Drug Saf 24:1113–1125
36. Freund B, Gravenstein S, Elliott M, Miller I (1999) Zanamivir: a review of clinical safety. Drug Saf 21:267–281
37. Hayden FG, Osterhaus AD, Treanor JJ (1997) Efficacy and safety of the neuraminidase inhibitor zanamivir in the treatment of influenza virus infections. GG167 Influenza Study Group. N Eng J Med 337:874–880
38. Makela MJ, Pauksens K, Rostila T (2000) Clinical efficacy and safety of the orally inhaled neuraminidase inhibitor zanamivir in the treatment of influenza: a randomized, double-blind, placebo-controlled European study. J Infect 40:42–48
39. Monto AS, Fleming DM, Henry D (1999) Efficacy and safety of the neuraminidase inhibitor zanamivir in the treatment of influenza A and B virus infections. J Infect Dis 180:254–261
40. Tanaka TNK, Murashima A (2009) Safety of neuraminidase inhibitors against novel influenza A (H1N1) in pregnant and breastfeeding women. CMAJ 181:55–58
41. Anonymous (2007) Monitoring of neuraminidase inhibitor resistance among clinical influenza virus isolates in Japan during the 2003–2006 influenza season. Wkly Epidemiol Rec 17:147–150
42. Wattanagoon Y, Stepniewske K, Lindergardh N, Pukrittayakamee S, Silachamroon U (2009) Pharmacokinetics of high-dose oseltamivir in healthy volunteers. Antimicrob Agents Chemother 53:945–952

43. Daniel MJ, Barnett JM, Pearson BA (1999) The low potential for drug interactions with zanamivir. Clin Pharmacokinet 36:41–50
44. Hayden FG (2001) Perspectives on antiviral use during pandemic influenza. Philos Trans R Soc Lond B Biol Sci 356:1877–1184
45. Moscona A (2005) Neuraminidase inhibitors for influenza. N Engl J Med 353:1363–1373
46. Jain S, Karimoto L, Bramley AM (2009) Hospitalized patients with 2009 H1N1 influenza in the United States, April-June 2009. N Engl J Med 361:1935–1943
47. Louie JK, Acosta M, Jamieson DJ, Honein MA, for the California Pandemic (H1N1) Working Group (2010) Severe 2009 H1N1 influenza in pregnant and postpartum women in California. N Engl J Med 362:27–35
48. World Health Organization. Clinical management of human infection with pandemic (H1N1) 2009: revised guidance, November 2009. Available at: http://www.who.int/csr/resources/publications/swineflu/clinical_management_h1n1.pdf. Accessed 10 Jan 2010
49. Centers for Disease Control and Prevention. Updated interim recommendations for the use of antiviral medications in the treatment and prevention of influenza for the 2009–2010 season. Available at: http://www.cdc.gov/h1n1flu/recommendations.htm#b. Accessed 01 Feb 2010
50. Hurt AC, Holien JK, Parker MW, Barr IG (2009) Oseltamivir resistance and the H274Y neuraminidase mutation in seasonal, pandemic and highly pathogenic influenza viruses. Drugs 68:2523–2531
51. Yamashita M, Tomozawa T, Kakuta M, Tokumitsu A, Nasu H, Kubo S (2009) CS-8958, a prodrug of the new neuraminidase inhibitor R-125489, show long-acting anti-influenza virus activity. Antimicrob Agents Chemother 53:186–192
52. Writing Committee of the Second World Health Organization Consultation (2008) Clinical aspects of human infection with avian influenza A (H5N1) virus. Update on avian influenza A (H5N1) virus infection in humans. N Engl J Med 358:261–273
53. Rameix-Welti MA, Agou F, Buchy P, Mardy S, Aubin JT, Veron M, van der Werf S, Naffakh N (2006) Natural variation can significantly alter the sensitivity of influenza A (H5N1) viruses to oseltamivir. Antimicrob Agents Chemother 50:3809–3815
54. Yen HL, Monto AS, Webster RG, Govorkova EA (2005) Virulence may determine the necessary duration and dosage of oseltamivir treatment for highly pathogenic A/Vietnam/1203/04 (H5N1) influenza virus in mice. J Infect Dis 192:665–672
55. Govorkova EA, Ilyushina NA, Boltz DA, Douglas A, Yilmaz N, Webster RG (2007) Efficacy of oseltamivir thaerapy in ferrets inoculated with different clades of H5N1 influenza virus. Antimicrob Agents Chemother 51:1414–1424
56. McKimm-Breschkin J, Selleck P, Usman TB, Johnson M (2007) Reduced sensitivity of influenza A (H5N1) to oseltamivir. Emerg Infect Dis 13:1354–1357
57. de Jong MD, Thanh TT, Khanh TH (2005) Oseltamivir resistance during treatment of influenza A (H5N1) infection. N Engl J Med 353:2667–2672
58. Saad MD, Boynton BR, Earhart KC (2007) Detection of oseltamivir resistance mutation N294S in humans with influenza A H5N1. In: Program and abstracts of the Options for the Control of Influenza Conference, Toronto, June 17–23, 228. abstract
59. World Health Organization. WHO rapid advice guidelines on pharmacological management of humans infected with avian influenza A (H5N1) virus. 2006. Available at: http://www.who.int/medicines/pubblications/WHO_PSM_PAR_2006.6.pdf. Accessed 20 Dec 2007
60. World Health Organization. Clinical management of human infection with avian influenza A (H5N1) virus. Available at http://www.who.int/csr/disease/avian_influenza/guidelines/clinicalmanagement07/pdf. Accessed 20 Dec 2007
61. Ilyushina NA, Hoffmann E, Solomon R, Webster RG, Govorkova EA (2007) Amantadine-oseltamivir combination therapy for H5N1 influenza virus infection in mice. Antivir Ther 12:363–370
62. Stittelaar KJ, Tisdale M, van Amerongen G, van Lavieren RF, Pistoor F, Simon J, Osterhaus AD (2008) Evaluation of intravenous zanamivir against experimental influenza A (H5N1) virus infection in cynomolgus macaques. Antiviral Res 80:225–228

63. Yun NE, Linde NS, Zacks MA, Barr IG, Hurt AC, Smith JN, Dziuba N, Holbrook MR, Zhang L, Kilpatrick JM (2008) Injectable peramivir mitigates disease and promotes survival in ferrets and mice infected with the highly virulent influenza virus, A/Vietnam/1203/04 (H5N1). Virology 374:198–209
64. Boltz DA, Ilyushina NA, Arnold CS, Babu YS, Webster RG, Govorkova EA (2008) Intramuscularly administered neuraminidase inhibitor peramivir is effective against lethal H5N1 influenza virus in mice. Antiviral Res 80:150–157
65. Hurt AC, Selleck P, Komadina N, Shaw R, Brown L, Barr IG (2007) Susceptibility of highly pathogenic A(H5N1) avian influenza viruses to the neuraminidase inhibitors and adamantanes. Antiviral Res 73:228–231
66. Cooper NJ, Sutton AJ, Abrams KR, Wailoo A, Turner D, Nicholson KG (2003) Effectiveness of neuraminidase inhibitors in treatment and prevention of influenza A and B: systematic review and meta-analyses of randomised controlled trials. BMJ 326:1235
67. Monto AS, Fleming DM, Henry D (1999) Efficacy and safety of the neuraminidase inhibitor zanamivir in the treatment of influenza A and B virus infections. J Infect Dis 180:254–261
68. Murphy KR, Eivindson A, Pauksens K (2000) Efficacy and safety of inhaled zanamivir for the treatment of influenza in patients with asthma or chronic obstructive pulmonary disease: a double-blind, randomized, placebo-controlled, multicentre study. Clin Drug Investig 20:337–349
69. Hedrick JA, Barzilai A, Behre U (2000) Zanamivir for treatment of symptomatic influenza A and B infection in children five to twelve years of age: a randomized controlled trial. Pediatr Infect Dis J 19:410–417
70. Aoki FY, Macleod MD, Paggiaro P (2003) Early administration of oral oseltamivir increases the benefits of influenza treatment. J Antimicrob Chemother 51:123–129
71. Kawai N, Ikematsu H, Iwaki N (2005) Factors influencing the effectiveness of oseltamivir and amantadine for the treatment of influenza: a multicenter study from Japan of the 2002–2003 influenza season. Clin Infect Dis 40:1309–1316
72. Burch JPM, Conti S for the CRD/CHE Technology Assessment group (Centre for Reviews and Dissemination/Centre for Health economics), University of York. Influenza – zanamivir, amantadine and oseltamivir (review): assessment report, 2008. Available at: http://www.nice.org.uk:Guidance/TA168. Accessed 5 Jan 2010
73. Whitley RJ, Hayden FG, Reisinger KS (2001) Oral oseltamivir treatment of influenza in children. Pediatr Infect Dis J 20:127–133 [Published correction appears in P*ediatr Infect Dis J*. 20:421]
74. Hayden FG, Pavia AT (2006) Antiviral management of seasonal and pandemic influenza. J Infect Dis 1:S119–126
75. Lalezari J, Campion K, Keene O, Silagy C (2001) Zanamivir for the treatment of influenza A and B infection in high risk patients. Arch Intern Med 161:212–217
76. Piedra PA, Schulman KL, Blumentalks WA (2009) Effects of oseltamivir on influenza-related complications in children with chronic medical complications. Pediatrics 124:170–178
77. McCullers JA, Bartmess KC (2003) Role of neuraminidase in lethal synergism between influenza virus and Streptococcus pneumoniae. J Infect Dis 187:1000–1009
78. Kaiser L, Wat C, Mills T (2003) Impact of oseltamivir treatment on influenza-related lower respiratory tract complications and hospitalizations. Arch Intern Med 163:1667–1672
79. Johnston SL, Ferrero F, Garcia ML, Dutkowski R (2005) Oral oseltamivir improves pulmonary function and reduces exacerbation frequency for influenza-infected children with asthma. Pediatr Infect Dis J 24:225–232
80. Hanshaoworakul W, Simmerman JM, Narueponjirakul U, Sanasuttipun W, Shinde V, Kaewachana S, Areechokechai D, Levy J, Ungchusak K (2009) Severe human influenza infections in Thailand: oseltamivir treatment and risk factors for fatal outcome. PLoS One 4: e6051
81. McGeer A, Green KA, Plevneshi A (2007) Antiviral therapy and out comes of influenza requiring hospitalization in Ontario, Canada. Clin Infect Dis 45:1568–1575

82. Lee N, Chan PK, Choi KW (2007) Factors associated with early hospital discharge of adult influenza patients. Antivir Ther 12:501–508
83. Sugaya N, Mitamura K, Yamazaki M, Tamuira D, Ichikawa M, Kimura K, Kawakami C, Kiso M, Ito M, Hatakeyama S (2007) Lower clinical effectiveness of oseltamivir against influenza B contrasted with influenza A infection in children. Clin Infect Dis 44:197–202
84. Sugaya N, Tamura D, Yamazaki M, Ichikawa M, Kawakami C, Kawaoka Y, Mitamura K (2008) Comparison of the clinical effectiveness of oseltamivir and zanamivir against influenza virus infection in children. Clin Infect Dis 47:339–345
85. Harper SA, Bradley JS, Englund JA, File TM, Gravenstein S, Hayden FG, McGeer AJ, Neuzil KM, Pavia AT, Tapper ML (2009) Seasonal influenza in adults and children-diagnosis, treatment, chemoprophylaxis, and institutional outbreak management: clinical practice guidelines of the Infectious Diseases Society of America. Clin Infect Dis 48:1003–1032
86. Monto AS, Pichichero ME, Blanckenberg SJ (2002) Zanamivir prophylaxis: an effective strategy for the prevention of influenza types A and B within households. J Infect Dis 186: 1582–1588
87. Hayden FG, Belshe R, Villanueva C (2004) Management of influenza in households: a prospective, randomized comparison of oseltamivir treatment with or without postexposure prophylaxis. J Infect Dis 189:440–9
88. Hayden FG, Gubareva LV, Monto AS (2000) Inhaled zanamivir for the prevention of influenza in families. Zanamivir Family Study Group. N Engl J Med 343:1282–1289
89. Welliver R, Monto AS, Carewicz O (2001) Effectiveness of oseltamivir in preventing influenza in household contracts: a randomized controlled trial. JAMA 285:748–754
90. Bowles SK, Lee W, Simor AE (2002) Use of oseltamivir during influenza outbreaks in Ontario nursing homes, 1999–2000. J Am Geriatr Soc 50:608–616
91. Schilling M, Povinelli L, Krause P (1998) Efficacy of zanamivir for chemoprophylaxis of nursing home influenza outbreaks. Vaccine 16:1771–1774
92. Peters PH Jr, Gravenstein S, Norwood P (2001) Long-term use of oseltamivir for the prophylaxis of influenza in a vaccinated frail older population. J Am Geriatr Soc 49:1025–1031
93. Monto AS, Rotthoff J, Teich E (2004) Detection and control of influenza outbreaks in well-vaccinated nursing home population. Clin Infect Dis 39:459–464
94. Vu D, Peck AJ, Nichols WG (2007) Safety and tolerability of oseltamivir prophylaxis in hematopoietic stem cell transplant recipients: a retrospective case-control study. Clin Infect Dis 45:187–193
95. Brokstad KA, Cox RJ, Olofsson J (1995) Parenteral influenza vaccination induces a rapid systemic and local immune response. J Infect Dis 171:198–203
96. Lawson F, Baker V, Au D (2000) Standing orders for influenza vaccination increased vaccination rates in inpatient settings compared with community reates. J Gerontol A Biol Sci Med Sci 55:M522–526
97. Tappenden PJR, Cooper K (2009) Amantadine, oseltamivir and zanamivir for the prophylaxis of influenza (including a review of existing guidance no. 67): systematic review and economic evaluation. Health Technol Assess 13:1–268

Resistance Development to Influenza Virus Sialidase Inhibitors

Marie-Anne Rameix-Welti, Sandie Munier, and Nadia Naffakh

Abstract The emergence of influenza viruses resistant to sialidase (neuraminidase, NA) inhibitors (NAIs) upon treatment is observed at low frequencies in adult immunocompetent patients, and does not impact significantly the evolution of influenza syndrome. In contrast, the development of resistance to NAIs is of clinical concern in immunocompromised patients and in patients with zoonotic influenza. Resistance mutations affect residues that are close to or belong to the catalytic site of the enzyme. Structural data provide clues about the subtype specificity of resistance mutations and the absence of cross-resistance to oseltamivir and zanamivir. Experimental data pointed to a negative impact of resistance mutations on viral fitness and transmissibility. However, H1N1 viruses naturally resistant to oseltamivir did emerge and spread in 2007–2008, which compromises the clinical effectiveness of the drug. This unexpected event stresses the need for a close surveillance of the impact of genetic drift on the susceptibility of influenza viruses to NAIs and for the further development of novel anti-influenza drugs.

M.-A. Rameix-Welti
Institut Pasteur, Unité de Génétique Moléculaire des Virus à ARN, Département de Virologie, F-75015, Paris, France

CNRS URA3015, F-75105, Paris, France

Université Paris-Diderot, Sorbonne Paris Cité, Unité de Génétique Moléculaire des Virus à ARN, F-75015, Paris, France

Université Versailles Saint-Quentin-en-Yvelines, F-78280, Guyancourt, France

S. Munier • N. Naffakh (✉)
Institut Pasteur, Unité de Génétique Moléculaire des Virus à ARN, Département de Virologie, F-75015, Paris, France

CNRS URA3015, F-75105, Paris, France

Université Paris-Diderot, Sorbonne Paris Cité, Unité de Génétique Moléculaire des Virus à ARN, F-75015, Paris, France
e-mail: nadia.naffakh@pasteur.fr

M. von Itzstein (ed.), *Influenza Virus Sialidase - A Drug Discovery Target*, Milestones in Drug Therapy, DOI 10.1007/978-3-7643-8927-7_8,

1 Introduction

The genetic variability of influenza viruses, driven by an elevated error rate of the viral RNA-dependent RNA polymerase during replication and by genetic reassortment, favors the rapid selection of drug-resistant variants. This matter of fact has considerably limited the usefulness of the adamantane-based M2 ion channel inhibitors, amantadine and rimantadine, in the treatment of influenza A infections. Both drugs have given rise to the emergence of drug-resistant viral strains within 2–5 days of treatment in up to 30% of patients. Drug-resistant viruses are genetically stable, show unaltered virulence and transmissibility compared to sensitive viruses, and have been isolated from patients who had not been exposed to an M2 inhibitor [1].

The sialidase inhibitors zanamivir and oseltamivir were developed based on the three-dimensional structure of the enzyme–sialic acid complex. They interact with amino acids in the active site of the viral sialidase that are highly conserved in all A and B strains. It was hoped that sialidase inhibitors would not rapidly give rise to viable resistant viruses, as functionally essential residues had been targeted. This hope was initially supported by the results of clinical and animal studies, but was recently contradicted by the emergence and worldwide circulation of influenza viruses showing natural resistance to oseltamivir.

2 Mechanisms of Resistance to Sialidase Inhibitors

Because the viral HA and NA have antagonistic functions, resistance to NAI can be conferred by two types of mechanisms which are non-mutually exclusive: (a) mutations on the NA that reduce the affinity of the enzyme for the inhibitor and (b) mutations on the HA that reduce viral binding to sialic acids, and thus reduce viral dependence on NA activity. As zanamivir and oseltamivir have different chemical structures, mutations on the NA may confer resistance to one, but not necessarily to both inhibitors. In contrast, mutations on the HA usually confer resistance to both. There are a number of reports of the selection of resistant mutants upon serial passages under increasing concentrations of the NAI in Madin Darby Canine Kidney (MDCK) cells, or upon clinical treatment with the NAI. The various resistance mutations on the HA and NA are not observed with the same frequencies in vitro and in vivo.

The earlier and predominant mutations generated in vitro map to the HA gene, suggesting that the selection of HA mutants is favored on cultured cells. These mutations affect residues that are close to or belong to the receptor-binding site and reduce the affinity of HA for the sialic acids [2], whereas the affinity of the NA for zanamivir and oseltamivir remains unchanged. They allow viral growth in the presence of concentrations of NAI that inhibit the replication of a wild-type virus. In contrast to that observed in vitro, only few HA mutations have been detected in vivo (Table 1) [12, 18, 20].

Table 1 NAI resistance mutations on the NA observed in patients upon antiviral treatment

NA mutation[a]	HA mutation[b]	n[c]	Viral subtype	Treatment	Immune status	Age	IC50 fold increase[d]		Ref
							Zanamivir	OC	
H274Y	T137 + G225[e]	1	A/H1N1	OC	IC	Adult	1–1.5	>1,500	[3]
	T137 + N225[e]	1		OC	IC	Adult	nd	>1,500	[3]
	nd	1		OC	IC	Child	nd	R	[4]
	nd	1		OC and Zanamivir	ID	Adult	S	R	[5, 6]
	nd	1		OC	ID	na	1–2	300	[7]
	nd	3		OC	IC	Children	nd	500	[8]
	Variations	1	A/H5N1	OC	IC	Child	1–2	>1,500	[9]
	nd	2		OC	IC	Child and Adult	nd	nd	[10]
N294S	nd	2	A/H5N1	OC	IC	na	nd	nd	[11]
N294S + H274Y	Variations	1		OC	IC	Child	4–6[f]	15[f]	[9]
N294S	no	1	A/H3N2	OC	IC	Child	nd	300	[12]
E119V	nd	1	A/H3N2	OC	IC	Child	nd	R	[4]
	nd	1		OC	ID	Adult	S	R	[5]
	no	1		OC	IC	Adult	nd	R	[13]
	no	2		OC	IC	Children	nd	500	[12]
	R142G + Y195F	1		OC	ID	Adult	1	110	[14]
	V226F	1		OC	ID	Adult	2.8	276	[14]
	nd	1		OC	IC	na	1	20	[7]
E119V (+I222V)	Variations	1		OC then Zanamivir	ID	Child	5–8	100–300	[15]
Deletion 245–248							3	175	[16]
R292K	nd	1	A/H3N2	OC	IC	Adult	nd	9,000	[17]
	no	3		OC	IC	Adults	nd	10,000	[13]
	S262N[g]	6		OC	IC	Children	nd	>10,000	[12]
	nd	8		OC	IC	Children	nd	R	[4]
	nd	1		OC	IC	Child	nd	500	[8]

(continued)

Table 1 (continued)

NA mutation[a]	HA mutation[b]	n[c]	Viral subtype	Treatment	Immune status	Age	IC50 fold increase[d]		Ref
							Zanamivir	OC	
R152K	T198I[h]	1	B	Zanamivir	ID	Child	1,000	50–200	[5, 18]
D198N	nd	1	B	OC	ID	Adult	S	R	[5]
	S285A	1		OC and Zanamivir	ID	Child	11	8	[14]
G402S	no	1	B	OC	IC	Child	7	4	[19]

[a]N2 numbering
[b]H3 numbering
[c]Number of cases reported
[d]*S*/*R* sensitive/resistant (IC50 fold increase not available)
[e]Egg adaptation mutations initially present in the HA of the strain used for challenge
[f]In the presence of the N294S substitution
[g]In one case out of six
[h]The T198I mutation results in the loss of a glycosylation site at residue 196
no no mutation found in the HA, *nd* not determined, *OC* Oseltamivir carboxylate, *IC* Immunocompetent, *ID* Immunodepressed

However, the contribution of these mutations to viral NAI resistance remains to be established. It is also unclear to what extent HAs with low affinity for sialic acids expressed on human respiratory cells, such as the HAs of waterfowl influenza A viruses, can confer decreased sensitivity to NAI upon viral transmission from birds to humans. In the particular case of A/Vietnam/1203/04 (H5N1) recombinant viruses, mutations in the HA which caused a switch from avian to human receptor specificity did not alter susceptibility to the NAI in vitro [21].

The resistance mutations selected in vitro and which map to the NA gene affect conserved residues of the NA that are close to (Glu119, His274) or belong to (Arg292) the catalytic site of the enzyme (Table 2) [22, 30–35].

Despite the emergence of viruses showing resistance to zanamivir in vitro, there has been a single case of in vivo zanamivir resistance reported so far. A zanamivir-resistant type B virus with an Arg to Lys mutation at the catalytic residue 152 of the NA was isolated from an immunocompromised child treated with zanamivir (Table 1) [18]. In contrast, several types of oseltamivir-resistant viruses, harboring mutations at the Arg292 catalytic residue or at the framework residues Glu119, His274, Asn294, Asp198, or Gly402, have been isolated from oseltamivir-treated patients (Table 1) [4–10, 12–19]. The levels of resistance reported for a given mutation can vary significantly from one study to another, depending on the methodology used for IC50 determination (Table 1). In a comparative study by Mishin et al. [37], the most commonly observed R292K, E119V, and H274Y mutations were found to increase the IC50 for oseltamivir about 5,000-, 200-, and 700-fold, respectively [37]. The R292K mutation is the only one to be associated with some level of resistance to zanamivir (Table 2). In contrast, the D198N and G402S mutations, observed exclusively on type B viruses, confer cross-resistance to both oseltamivir and zanamivir, and have a moderate 5- to 10-fold effect on the IC50 (Table 1) [14, 19].

The level of resistance conferred by a given mutation is dependent on the genetic background. The R292K and E119V mutant viruses isolated in vivo were exclusively of the H3N2 subtype (Table 1). Recombinant type B viruses with these mutations were produced by reverse genetics, and showed the same resistant phenotype as the H3N2 viruses (Table 2) [27]. But attempts to rescue recombinant A/WSN/33 viruses with the R292K or E119V mutations were unsuccessful, suggesting a lethal effect in the H1N1 background [23, 24]. On the contrary, the H274Y mutant viruses isolated in vivo were exclusively of the H1N1 or H5N1 subtypes (Table 1), and recombinant H3N2 viruses with the H274Y mutation showed unaltered sensitivity to oseltamivir (Table 2) [26]. As mentioned earlier, the D198N and G402S mutations are specific to type B viruses. A notable exception is the N294S mutation, found both on a H3N2 virus and on a H5N1 virus, associated with a 300-fold resistance to oseltamivir [12], and a 15-fold resistance to oseltamivir and a 4- to 6-fold resistance to zanamivir, respectively [9] (Table 1).

The subtype specificity of resistance mutations results from distinct structural features of group 1 (N1, N4, N5, and N8) and group 2 (N2, N3, N6, N7, and N9) sialidases. Tight binding of oseltamivir carboxylate to the wild-type NA requires reorientation of the side chain of residue Glu276, which creates a hydrophobic

Table 2 NAI resistance mutations on the NA selected in vitro or studied by reverse genetics

NA mutation[a]	Experimental system[b]	Selective drug	NA subtype	NA activity[c] (%)	Sensitivity to NAI[d] (IC50 fold increase)		Ref
					Zanamivir	OC	
H274Y	In vitro	Oseltamivir	N1	50%	S (nd)	R (300)	[22]
	RG	na	N1	28–88%	S (1)	R (292–1,672)	[23–25]
	RG	na	N2	37%	S (1.2)	S (0.8)	[26]
N294S	RG	na	N1	33–88%	RS (3)	R (21–113)	[24, 25]
	RG	na	B	77%	RS (4.7)	R (252)	[27]
E119V	RG	na	N2	28–81%	S (0.8)	R (8.5–260)	[26, 28, 29]
	RG	na	B	142%	S (1.9)	R (>300)	[27]
E119V (+I222V)	RG	na	N2	nd	nd	R (224)	[29]
E119G	In vitro	Zanamivir	N2	46%	R (700)	nd	[30, 31]
	In vitro	Zanamivir	N9	nd	R (250)	nd	[32, 33]
	In vitro	Zanamivir	B	nd	R (500)	nd	[33, 34]
	RG	na	B	72%	R (>560)	RS (31.1)	[27]
E119A/D	In vitro	Zanamivir	N2	23/3%	R (600)/R (2,500)	nd	[30, 31]
	RG	na	N2	nd/19%	nd/R (32)	nd/S (1.6)	[26]
	RG	na	B	174%/29%	R/R (>560)	R/R (>300)	[27]
R292K	In vitro	Zanamivir	N2	44%	RS (10)	nd	[31]
	In vitro	Oseltamivir	N2	<10%	RS (24)	R (30,000)	[35]
	RG	na	N2	6–29%	RS (2.5–13)	R (>1,580– > 30,000)	[26, 28, 36]
	RG	na	B	6%	RS (28.5)	R (>300)	[27]
R152K	RG	na	N2	1.7%	S (1)	S (1)	[36]
D151E	RG	na	N2	2.2%	S (2)	RS (11)	[36]
R371K	RG	na	N2	2.6%	R (15)	R (45)	[36]

[a]N2 numbering

[b]In vitro: mutant virus selected in the presence of the NAI, RG: mutant virus rescued by reverse genetics (only mutations that have been associated with drug resistance in patients are reported here)

[c]Residual sialidase activity of the mutant virus compared to the wild type

[d]*R* resistant, *S* sensitive, *RS* reduced susceptibility, as stated by the authors

na not applicable, *nd* not determined, *OC* Oseltamivir carboxylate

pocket to accommodate the *O*-ethyl-propyl group of the inhibitor. The crystal structure of a H274Y mutant N1 in complex with oseltamivir showed that in the presence of the bulkier Tyr274 residue, the required conformational change of Glu276 is blocked and the hydrophobic pocket is not able to form, resulting in a weak binding of oseltamivir (Fig. 1a) [40].

Group 2 NAs can accommodate the H274Y substitution without disrupting the oseltamivir-binding site because they have a Thr residue at position 252 beneath residue 274, instead of the bulkier Tyr252 residue present in group 1 NAs [38, 40]. In group 2 NAs, the R292K substitution confers resistance to oseltamivir most likely due to a disruption of a hydrogen bond between the Arg292 and the carboxylate moiety of oseltamivir, and favors a Lys292-Glu276 electrostatic interaction which prevents reorientation of the side chain of residue Glu276 (Fig. 1b) [39]. In group 1 NAs, the R292K substitution is less detrimental, probably because of the presence of a conserved Tyr347 residue which makes a hydrogen bond with the carboxylate moiety of oseltamivir [38]. Such a bond is not made by the 347 residue found in group 2 NAs.

Structural data also provide clues about the absence of cross-resistance to oseltamivir and zanamivir. Zanamivir (as well as the sialic acid-based natural substrate) has a polar glycerol group at the C6 position, instead of the hydrophobic *O*-ethyl-propyl group characteristic of oseltamivir. Thus, binding of zanamivir does not require reorientation of Glu276, but rather involves the formation of a hydrogen bond between Glu276 and the glycerol group. As a consequence, the H274Y and R292K substitutions that block reorientation of Glu276 in group 1 and group 2 NAs, respectively, do not prevent tight binding of zanamivir [39, 40]. This chemical property of zanamivir probably contributes to the low frequency of resistance observed in treated patients. Interestingly, the structure of a N294S mutant N1 in complex with oseltamivir suggests that the loss of the Asn side chain, which points toward Tyr347 in the wild-type NA, leads to an increased flexibility of the side chain of Tyr347 and a weaker hydrogen bond with the carboxylate group of oseltamivir. The presence of a carboxylate group at the C2 position is a shared feature of oseltamivir, zanamivir, and sialic acid, consistent with the fact that the N294S mutant N1 showed weaker binding for all three compounds compared to the wild type [39, 40].

3 Impact of Resistance to Sialidase Inhibitors on Viral Fitness

Resistance mutations affect the enzymatic properties of the NA. The R292K mutation in an H3N2 background was shown to reduce the affinity of the NA for a sialic acid-like synthetic substrate 4- to 40-fold, as indicated by the increase in K_m values [35, 41, 42], and to decrease its enzymatic activity (V_{max}) [35]. Similarly, the H274Y substitution was shown to reduce the affinity of the NA of H1N1 and H5N1 viruses for a sialic acid-like substrate and/or the NA enzymatic activity [22, 24, 43]. In addition, the R292K, R152K, and E119G/V mutations may decrease the

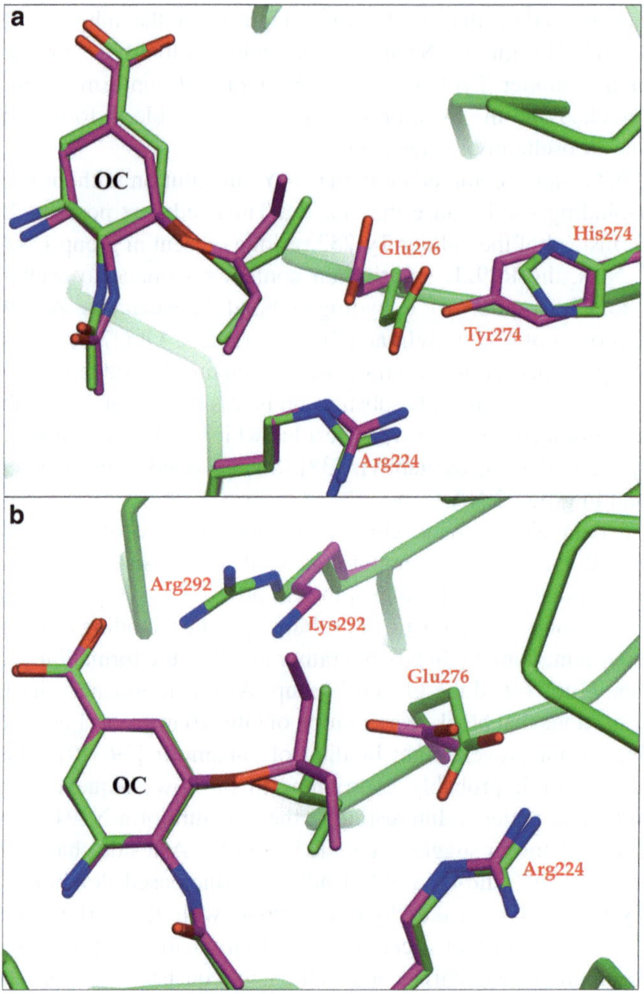

Fig. 1 Crystal structure of the active sites of a group 1 and a group 2 sialidase bound to oseltamivir carboxylate, in the absence or in the presence of a mutation conferring resistance to oseltamivir. The crystal structures of the A/Vietnam/1203/04 (H5N1) (**a**) and A/Tern/Australia/G70C/75 (H11N9) (**b**) sialidase active sites bound to oseltamivir carboxylate are represented. Carbons of the wild-type N1 (PDB #2HU4) [38] and N9 (PDB #2QWK) [39] structures are represented in green. Carbons of the H274Y mutant N1 (PDB #3CL0) [40] and R292K mutant N9 (PDB #2QWH) [39] are represented in magenta. The position of oseltamivir carboxylate (OC) in the active site of the wild-type and mutant NAs is represented in green and in magenta, respectively. In the wild-type N1 and N9 (**a** and **b**), reorientation of the side chain of residue Glu276 is stabilized by an electrostatic interaction with residue Arg224, which lets a hydrophobic pocket accommodate the O-ethyl-propyl group substituted on C6 of the inhibitor. In the H274Y mutant N1 (**a**), the replacement of the His274 by the bulkier Tyr274 residue blocks the reorientation of Glu276. In the R292K mutant N9 (**b**), the reorientation of the Glu276 side chain is prevented by an electrostatic interaction with the Lys292 residue. As a result, the hydrophobic pocket that is needed for tight OC binding is disrupted in both mutant NAs. Figures were kindly provided by Jeffrey Dyason and Mark von Itzstein

thermostability of the enzyme [27, 28, 35, 44]. Such alterations of NA properties are likely to affect the viral replicative potential.

The impact of NA resistance mutations on viral fitness was documented in both cultured cells and animal models. Recombinant H3N2 viruses carrying the mutation E119V showed the same kinetics of multiplication compared to the wild type in cultured cells [26, 28]. In contrast, H3N2 viruses with the R292K mutation, either recombinant or selected in vitro, showed a profound replication defect in MDCK or MDCK-SIAT1 cells compared to the wild type [26, 28, 35]. A H1N1 oseltamivir-resistant clinical isolate, and recombinant or in vitro-selected A/WSN/33 viruses that carry the H274Y mutation showed compromised replication efficiency compared to their wild-type counterparts on MDCK cells [23, 45, 46]. However, some R292K mutant viruses [31, 41] and a H274Y mutant A/PR/8/34 virus [24] were reported to grow as well as the wild type on MDCK cells. Overall, the in vitro replicative potential of oseltamivir-resistant viruses appears to be decreased, although to a variable extent, depending on the viral genetic background and on the cellular system used.

Mouse and ferret models were used to assess the virulence of NAI-resistant H3N2 or H1N1 isolates with the R292K, E119V, or H274Y mutations. In mice, all three mutations significantly compromised viral infectivity and/or replication ability, as indicated by the reduction in viral titers in the lungs at day 3 postinfection (1 to 2 logs for the E119V, 2 logs for the R292K, and 3 logs for the H274Y mutations) [41, 46, 47]. Reduced virulence in mice was also observed for in vitro-selected H3N2- and H1N1-resistant viruses [2, 31, 35] and, although to a lesser extent, for recombinant A/WSN/33 and A/PR/8/34 strains with the H274Y mutation [24, 48]. In the upper respiratory tract of ferrets, oseltamivir-resistant H1N1 and H3N2 isolates [41, 46, 47], as well as an oseltamivir-resistant H5N1 human isolate showing the H274Y mutation [9], replicated less efficiently than their wild-type counterparts. However, in the background of a recombinant virus derived from the A/Vietnam/1203/04 human H5N1 isolate, neither the H274Y nor the N294S mutation compromised viral pathogenicity and lethality in mice [24]. This observation suggests that viruses with a high replicative potential may accommodate mutations of resistance on the NA with no decrease in virulence.

The transmissibility of NAI-resistant viruses was examined in the ferret and the guinea pig models. A H274Y mutant H1N1 virus showed delayed transmission in a contact ferret model compared to the wild type, although a 100-fold higher dose of the mutant virus had been inoculated to donor ferrets [49]. Even more strinkingly, a H3N2 clinical isolate or a recombinant H3N2 virus carrying the R292K mutation was not transmitted in a contact ferret model under conditions that allowed a 100% efficient transmission of their wild-type counterparts [28, 50]. In contrast, recombinant H3N2 viruses with the E119V mutation were transmitted as efficiently as the wild-type virus into a contact ferret model [28] or a contact guinea pig model [29]. Overall, the transmissibility of NAI-resistant viruses seems to correlate with the degree of functional alteration of NA enzymatic properties. Noticeably, however, a E119V mutant virus transmitted only poorly when the guinea pig was used as a model for aerosol rather than contact transmission [29].

Experimental data suggested that H1N1 viruses with a H274Y mutation in the NA had such a decreased fitness and transmissibility that their diffusion in the human population was very unlikely. However, H1N1 viruses naturally resistant to oseltamivir did emerge and spread in 2007–2008. Although resistance was linked to the H274Y mutation, resistant and sensitive H1N1 viruses showed similar growth properties in vitro [51]. Enzymatic analysis suggested that a specific combination of amino acids resulted in an increased affinity of the N1 of recent H1N1 viruses for sialic acids, and that an appropriate functional balance between the activities of the HA and NA toward sialic acids was restored by the H274Y mutation on the NA [51]. These observations underline the fact that genetic variations of the hemagglutinin and sialidase, which are mainly driven by the immune response, can lead to the co-selection of resistance mutations and adventitious mutations that result in increased fitness and transmissibility.

4 Monitoring Resistance to Sialidase Inhibitors

Both phenotypic and genotypic assays can be used for the detection of NAI-resistant viruses. Viruses with resistance mutations in the NA, but not in the HA, are readily detected by using an in vitro enzyme inhibition assay to determine the IC50 (concentration of inhibitor resulting in a 50% reduction in the sialidase activity, or Inhibiting Concentration 50). Cell culture assays, which allow the determination of EC50 values (concentration of inhibitor resulting in a 50% reduction in the viral yield, or Effective Concentration 50), account for both NA- and HA-mediated resistance. Both assays are thus useful and complementary.

The NA inhibition assay is based on a fluorometric NA activity assay [52] which was further optimized for IC50 determination [53]. The fluorescence of the 4-methylumbelliferone released upon NA-mediated cleavage of the fluorogenic substrate 2′-(4-methylumbelliferyl)-α-D-*N*-acetylneuraminic acid (MUNANA) is measured in the absence or in the presence of antiviral drugs. A more sensitive assay was developed, using a chemiluminescent substrate (1,2-dioxetane derivative of sialic acid, NA-STAR), [54] instead of the MUNANA. The IC50 values show some variability from one study to another, depending on whether the fluorogenic or chemiluminescent substrate is used and on the substrate concentration that is used [55]. Both substrates have also been used to determine the affinity of the NA for zanamivir and oseltamivir (Ki) [22, 41, 43, 56].

The evaluation of viral sensitivity to NAI in cell culture assays is usually based on the degree of inhibition of virus yield upon multicycle growth (as measured by hemagglutination or pfu titers, or by TCID50), viral plaque formation (as determined by the number and size of plaque-forming units), or viral infectivity (as determined by a FACS assay following immunohistochemical staining with an anti-NP antibody) [57]. Such assays were initially developed in the standard MDCK cell line. A poor correlation was found between the sensitivity of clinical isolates to NAIs in cell culture and their sensitivity in the enzyme inhibition assay

[58, 59]. The fact that sensitive isolates generally behaved as drug resistant in cell culture assays was probably due to the low amounts of sialyl-α2,6-galactose-containing receptors on the surface of MDCK cells, which reduced viral dependency on NA sialidase activity. Conversely, a drug-resistant isolate with the R152K mutation appeared as drug sensitive on MDCK cells because of the presence of a mutation in the HA, which increased viral binding to sialyl-α2,3-galactose-containing receptors on the surface of MDCK cells, and thus increased viral dependency on NA sialidase activity [18]. Assays performed on MDCK cells overexpressing the human α2,6-sialyltransferase (MDCK-SIAT1 cells) proved to be a more suitable system to monitor sensitivity of human viruses to NAIs [60, 61]. Primary normal human bronchial epithelial cells, which are functionally and morphologically close to normal differentiated human airway epithelium, may provide an optimal system for prediction of virus resistance in vivo [21, 62].

Resistance mutations present in the NA gene are classically detected by viral RNA extraction, RT-PCR, and sequencing of the resulting amplicon. In addition to this accurate but time-consuming method, novel approaches are being developed for the rapid detection of well-characterized resistance mutations. For instance, a one-step allelic discrimination RT-PCR assay using two distinctly labeled probes specific for the wild-type His274 and the mutant Tyr274 codon allowed the rapid detection of the H274Y substitution in circulating H1N1 viruses [63] and in highly pathogenic avian influenza H5N1 [64]. Pyrosequencing protocols were recently developed for rapid detection of the most commonly reported mutations associated with NAI resistance in seasonal H1N1 and H3N2, and in H5N1 viruses [65–68]. Interestingly, pyrosequencing enabled accurate quantification of the relative proportions of resistant cultured viruses when experimentally mixed with sensitive cultured viruses and when present in the total virus population at levels as low as 10% [65, 67, 68].

5 Incidence of Resistance to Sialidase Inhibitors Under Treatment

In immunocompetent patients, the emergence of seasonal influenza viruses resistant to NAIs in clinical trials was observed at much lower frequency compared to the emergence of amantadine-resistant viruses [1, 69]. No zanamivir-resistant virus has been isolated from immunocompetent patients treated with zanamivir so far, which could be partly due to the infrequent usage of zanamivir in the clinic and in clinical trials compared to oseltamivir. Viruses with a decreased sensitivity to oseltamivir associated with a H274Y mutation were detected in up to 4% of adult volunteers who had been experimentally infected with H1N1 viruses and subsequently treated with oseltamivir [3]. Based on cumulative data from Roche-sponsored clinical trials involving more than 1,000 adults and 500 children, the incidence of viruses with reduced susceptibility to oseltamivir is about 0.32–0.4% in adults and 4–5% in children infected with influenza A viruses [70].

The higher frequency of emergence of oseltamivir-resistant viruses in children could be related to a higher viral replication, a prolonged viral shedding, and an increased clearance of the drug compared to that in adults [71]. These findings led to the recommendation of a weight-dependent dosage of oseltamivir in children [72]. Two Japanese studies reported the isolation of drug-resistant viruses in 5.5% and 16–18% of children treated with the initial 2 mg/kg dosage following infection with H1N1 and H3N2 viruses, respectively. A reduced incidence rate of 1.1% was reported in recent studies when the weight-dependent dosage was used [70]. In two independent studies, the incidence of resistant influenza B viruses in children treated with oseltamivir was found to be <1% [4, 19]. Notably, however, influenza B viruses are naturally less sensitive to oseltamivir compared with influenza A viruses (Table 3), and the current oseltamivir dosage may be suboptimal with regard to the emergence of resistant influenza B viruses.

In immunocompromised patients under treatment with NAIs, the emergence of viruses showing reduced drug susceptibility has been reported in at least nine cases (Table 4). The only reported case of emergence of a zanamivir-resistant virus upon treatment occurred in an immunocompromised child infected with influenza B virus (Table 4).

Table 3 IC50s to zanamivir and oseltamivir for seasonal influenza A and B viruses

Mean IC50 (nM)									Period	Area	Assay[c]	Ref
N1			N2[a]			B						
Zana	OC	n[b]	Zana	OC	n[b]	Zana	OC	n[b]				
0.76	1.2	139	1.82	0.5	767	2.28	8.8	148	1996–1999	World wide	F	[73]
1.14	0.9	4	2.09	0.73	38	4.15	11.53	23	1999–2000	Canada	L	[74]
0.37	0.66	235	1.04	0.31	169	1.4	14.84	128	1998–2002	Australia, New Zealand, South-east Asia	F	[75]
1.03	1.32	233[d]	3.05	0.65	482[d]	8.3	13.2	335[d]	2000–2002	World wide	L	[76]
0.40	0.51	622	1.38	0.37	1,119	1.86	5.37	743	1999–2002	World wide	L	[77]
0.92	1.34	9	2.51	0.73	178	4.19	12.99	80	2002–2003	France	F	[78]
nd	nd	nd	2.06	0.54	355[e]	nd	nd	nd	2002–2005	France	F	[79]
1.17	1.66	150	0.85	0.34	8	9.82	21.29	224	2006	France	F	[80]
nd	nd	nd	nd	nd	nd	10.1	70.5	422	2004–2005	Japan	F	[19]
1.06	0.91	1,128[f]	2.54	0.44	1,236[f]	3.87	3.42	889	2004–2007	World wide	L	[7]
0.36	0.67	288	1.02	0.32	540	1.88	13.06	270	2001–2006	Australia, South-east Asia, South pacific, Eastern Asia, South Africa	F	[81]

[a]H3N2 and H1N2 viruses

[b]Number of viruses assayed for resistance to oseltamivir and zanamivir

[c]NA inhibition assays were performed using either the MUNANA fluorogenic substrate (F) or the NA-STAR chemiluminescent substrate (L)

[d]Number of viruses assayed for resistance to oseltamivir. In this same study, 271 H1N1, 566 HxN2, and 712 B viruses were assayed for resistance to zanamivir.

[e]Including viruses from 2002 to 2003 reported in [78]

[f]Number of viruses assayed for resistance to oseltamivir. In this same study, 1,134 H1N1 and 1,233 H3N2 viruses were assayed for resistance to zanamivir.

nd not determined, *Zana* Zanamivir, *OC* oseltamivir carboxylate

Table 4 NAI-resistant viruses isolated from immunocompromised patients

NA	Mutation		Treatment	Age	Clinical context[c]	Issue	Sensitivity to NAI		Ref
	NA[a]	HA[b]					Zanamivir	OC	
N1	H274Y		Oseltamivir + Zanamivir	Adult	BMT	Death	S	R	[5, 6]
N1	H274Y		Pretreatment mutation	Adult	CT	Death	S	R	[82]
N2	E119V		Oseltamivir	Adult	na	na	S	R	[5]
N2	E119V		Oseltamivir	Adult	CT	Recovery	S	R	[14]
N2	E119V	V226F	Oseltamivir	Adult	Kidney graft	Death	S	R	[14]
N2	E119V (+I222V)		Oseltamivir, then Zanamivir	Child	BMT	Recovery	S	R	[15]
	Del 245-248						S	R	[16]
B	R152K	T198I[d]	Zanamivir	Child	BMT	Death	R	R	[18]
B	D198N		Oseltamivir	Adult	na	na	S	R	[5]
B	D198N	S285A	Oseltamivir + Zanamivir	Child	BMT	Death	S	R	[14]

[a]N2 numbering
[b]Mutation on the HA (H3 numbering) found in association with the resistance mutation on the NA
[c]*BMT* bone marrow transplantation, *CT* anticancerous chemotherapy
[d]The T198I mutation results in the loss of a glycosylation site at residue 196
na not available, *S* sensitive, *R* resistant, *OC* oseltamivir carboxylate

Persistent viral shedding during 12 and 17 months was reported in an immunocompromised child [15] and in an immunocompromised adult [6], respectively. It was suggested that such a prolonged viral shedding may favor the selection of resistant viruses [83]. However, given the limited number of case reports, the incidence of viruses resistant to NAIs in immunocompromised patients remains unknown.

Emergence of oseltamivir-resistant viruses during treatment was recently described in three patients infected with a highly pathogenic H5N1 virus [9, 10]. When sensitivity to oseltamivir was systematically monitored in eight patients during treatment, the emergence of a resistant H5N1 virus was observed in two of them (25%) [10]. These observations suggest that the incidence of resistant virus could be higher upon infection with H5N1 viruses compared to seasonal viruses. The difference could be due to higher viral loads and to a prolonged viral shedding due to the lack of previous immunity [84].

6 Incidence of Resistance to Sialidase Inhibitors in the Absence of Treatment

The emergence of resistant viruses that are able to spread in the human population is a public health concern, as it compromises the effectiveness of the drug. The dramatic increase in the proportion of H3N2 viruses naturally resistant to amantadine worldwide from 1–2% in 2002 to 90.5% in 2005–2006 [69] stressed the need for a global surveillance of viruses resistant to NAIs, and in particular to oseltamivir as the drug became widely used in some countries.

Notably, there is no widely accepted definition of resistance to NAIs. A unique IC50 threshold value cannot be easily defined, because IC50 values may vary depending on the experimental conditions used for determination, and may vary from one type/subtype of influenza virus to another when determined in parallel. A greater sensitivity to oseltamivir compared to zanamivir was generally observed for viruses of the H3N2 subtype, whereas the opposite trend was observed for H1N1 viruses (Table 3). Influenza B viruses are generally found to be less susceptible to NAIs than influenza A viruses, as indicated by 2–10-fold and 5–50-fold increased IC50 values for zanamivir and oseltamivir, respectively (Table 3). In surveillance studies, statistical analysis of the data obtained on a large panel of viruses has been used to define the mean IC50 of sensitive viruses within a given NA type/subtype [7, 73, 76, 80, 85]. For example, the worldwide survey performed by the Global Neuraminidase Inhibitor Susceptibility Network from 1999 to 2002 found that the mean IC50 values for oseltamivir and zanamivir were 0.37 and 1.38 nM, respectively, for H3N2 viruses ($n = 1{,}119$), 0.51 and 0.40 nM, respectively, for H1N1 viruses ($n = 622$), and 1.86 and 5.37 nM, respectively, for B viruses ($n = 743$) (Table 3) [85]. Viruses are defined as outliers if they show an IC50 value higher than the mean IC50 + 2SD or the mean IC50 + 3SD, depending on the study.

Among the outliers group, viruses with an IC50 value at least 10-fold higher compared to the mean IC50 of sensitive viruses are considered as resistant viruses if their NA gene contains a mutation already known to be associated with drug resistance; if not, they are considered as extreme outliers [7, 73, 77]. The NA mutations found in extreme outliers are of interest since most of them affect conserved residues located close to the catalytic site, e.g., residue 222 for type B viruses [19, 77] or residue 151 for H3N2 viruses [7, 73]. Even more striking, a type B virus isolated in the absence of treatment showed the R371K mutation [7]. Residue 371 is part of the Arg triad (Arg118, Arg292, and Arg371) shared by all influenza NAs together with bacterial sialidases [86]. The R371K mutant type B virus showed a stronger decrease in sensitivity to zanamivir and oseltamivir (40- and 500-fold, respectively [7]) than some viruses defined as truly resistant. This observation underlines how difficult it is to figure out the boundary between natural variations in the susceptibility to NAIs and true resistance.

Two worldwide surveillance studies performed through the 1996–1999 and 2004–2007 periods reported 3/1,054 (0.28%) and 10/3,253 (0.3%) extreme outliers, respectively [7, 73], suggesting that the frequency of extreme outliers remains stable. In contrast, the number of oseltamivir-resistant viruses isolated from untreated patients increased in the recent years. Noteworthy, H1N1 viruses with the H274Y mutation (7 in total) and B viruses with the D198N mutation (3 in total) were the only kind of oseltamivir-resistant viruses isolated from untreated patients from 1999 to 2007, suggesting that these two mutations are the less costly in terms of viral fitness [7, 19, 80, 85].

During the 2007–2008 influenza season, H1N1 viruses naturally resistant to oseltamivir emerged and spread in the human population. Resistance was conferred by the H274Y mutation of the NA. Resistant viruses were more prevalent in Europe (25%) than in the Americas (16%) or the Western Pacific region (5%) (Fig. 2) (http://www.who.int/csr/disease/influenza/ResistanceTable200806013.pdf).

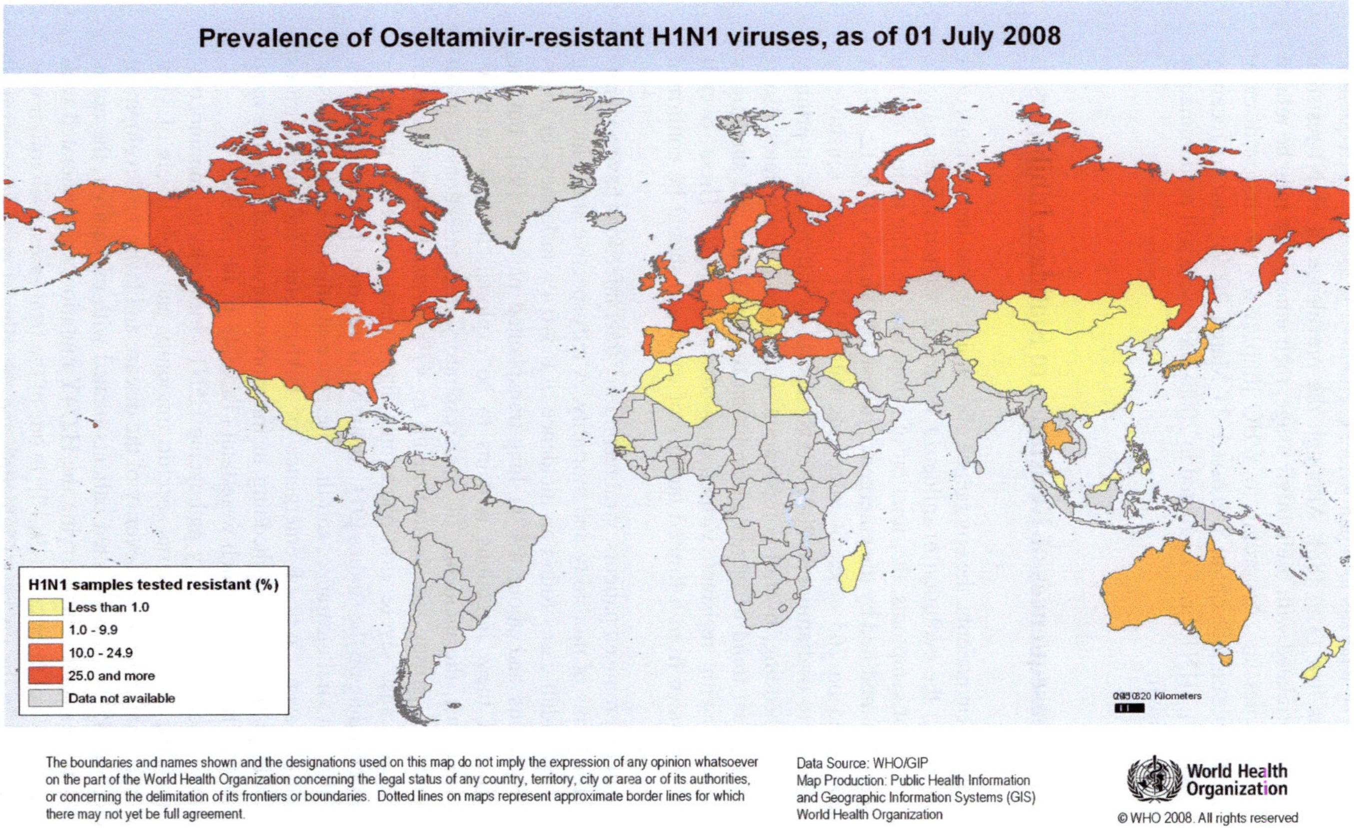

Fig. 2 Worldwide prevalence of oseltamivir-resistant H1N1 viruses, as of 01 July 2008. A color code is used to represent the percentage of oseltamivir-resistant H1N1 viruses. *Yellow*: < 1.0%; *light orange*: 1.0–9.9%; *dark orange*: 10.0–24.9%; *red*: > 25%; *gray*: data not available. Data source: WHO/GIP, Map Production Public Health Information and Geographic Information Systems (GIS), http://www.who.int/csr/disease/influenza/Global_H5N1Resistance_20080701.jpg With authorization

The highest prevalence was observed in Norway where oseltamivir usage is moderate (67%) [87], whereas in Japan where oseltamivir is used most frequently, the prevalence was low (3%) [88]. Although the prevalence of H1N1-resistant viruses was not correlated with oseltamivir usage, their emergence may be related to selective drug pressure. According to WHO [http://www.who.int/csr/disease/influenza/h1n1_table/en/index.html (accessed 23 March 2009)] and to recently published data [89], H1N1 viruses isolated in 2008–2009 were still predominantly resistant to oseltamivir.

7 Clinical Consequences of Resistance to Sialidase Inhibitors

In immunocompetent patients, the emergence of oseltamivir-resistant variants does not seem to impact the evolution of influenza syndrome. Cohort studies indicate that the intensity of symptoms, the duration of symptoms, and the course of viral excretion were not modified by the emergence of resistant viruses [4, 8, 12, 19]. A recent study, based on 265 cases of infections with H1N1 viruses sensitive or naturally resistant to oseltamivir in 2007–2008, reports no differences in primary symptoms, viral shedding, overall complication, and hospitalization rates. However, an insignificant trend of increased evolution to pneumonia and sinusitis in patients infected with resistant viruses was observed [87]. These overall conclusions are drawn from a limited number of cases and should be confirmed in the future.

In immunocompromised patients, viral respiratory tract infections have a significant impact in terms of morbidity and mortality. A retrospective study including more than 3,500 adult cases found an incidence of 0.3–0.4% in kidney- or liver-transplanted patients and about 4% in lung-transplanted patients [90], but the clinical impact of these infections remains to be explored. In bone marrow-transplanted patients, the incidence of viral respiratory tract infections is about 1%, and influenza viruses are isolated in up to 47% cases. In these patients, influenza viruses cause severe viral pneumonia, and mortality rates in the range of 4.5–47% depending on the studies [91]. An antiviral treatment is needed in order to achieve efficient viral clearance, and the emergence of antiviral resistance is of major concern. Seven cases of emergence of NAI-resistant influenza viruses associated with prolonged viral shedding in immunocompromised patients have been described so far, five of which were fatal (Table 4) [6, 14, 15, 18].

In patients infected with highly pathogenic H5N1 viruses, as in immunocompromised patients, lack of antivirus-specific immunity and high levels of viral replication may contribute to the severity of the disease and favor the development of resistance. In H5N1 virus-infected patients treated with oseltamivir, the emergence of a drug-resistant virus carrying the H274Y mutation was associated with fatal outcome in two of three cases [9, 10]. In one of the fatal cases, oseltamivir was administered early in the course of infection and the clinical condition remained stable during the first three days of treatment. The later isolation of an oseltamivir-resistant virus coincided with clinical deterioration and increased viral loads, suggesting that the development of drug resistance was responsible for treatment failure [10].

Overall, these observations highlight the need for additional antiviral agents that could be used instead of or in combination with oseltamivir for the treatment of H5N1 infections and for the treatment of immunocompromised patients.

8 Conclusion

Sialidase inhibitors are currently the drug of choice for the treatment of seasonal influenza, and their stockpiling is a key element of the pandemic preparedness plans in several countries. The development of viral resistance to NAIs is of clinical concern on an individual level in immunocompromised patients and in patients with zoonotic influenza. It becomes of concern at a population level as soon as resistant viruses are efficiently transmitted from human to human and are able to spread, because the clinical effectiveness of the drug is compromised overall. The emergence and circulation of influenza H1N1 viruses showing natural resistance to oseltamivir during the 2007–2008 season were an unexpected event, because clinical trials had resulted in low frequencies of oseltamivir resistance in adults, and animal studies had pointed to a negative impact of resistance mutation on viral fitness and transmissibility. The genetic and enzymatic characterization of these recent H1N1 viruses suggests that drift mutations on the NA contribute to the absence of debilitating effect of the H274Y resistance mutation. These findings stress the need for a better understanding of the interplay between resistance mutations and drift and/or compensatory mutations, on the NA, the HA, or possibly on other genomic segments, for all subtypes of circulating viruses. The impact of genetic drift on natural susceptibility of influenza viruses to NAIs, in the absence of a resistance mutation, also needs to be explored through further clinical, in vitro, and animal studies.

Strategies for antiviral resistance containment and surveillance still have to be improved. Enhanced methods for the rapid and accurate detection of resistant viruses are requested, as well as increased coverage and testing capacity of surveillance networks, and reinforced vigilance in the therapeutic usage of NAIs. The further development of novel anti-influenza drugs, targeting either the NA or other viral proteins such as the polymerase or the NS1 protein, is essential. Such developments would offer an alternative if circulating viruses become predominantly resistant to currently available compounds. They would also open the way to combination therapies, which are likely to reduce the emergence of drug-resistant viruses in at-risk populations such as children, immunocompromised, and H5N1 patients.

Acknowledgments We are very grateful to Jeffrey Dyason and Mark von Itzstein for providing Fig. 1.

References

1. Hayden FG, Hay AJ (1992) Emergence and transmission of influenza A viruses resistant to amantadine and rimantadine. Curr Top Microbiol Immunol 176:119–130
2. McKimm-Breschkin JL (2000) Resistance of influenza viruses to neuraminidase inhibitors-a review. Antiviral Res 47:1–17
3. Gubareva LV, Kaiser L, Matrosovich MN, Soo-Hoo Y, Hayden FG (2001) Selection of influenza virus mutants in experimentally infected volunteers treated with oseltamivir. J Infect Dis 183:523–531
4. Whitley RJ, Hayden FG, Reisinger KS, Young N, Dutkowski R, Ipe D, Mills RG, Ward P (2001) Oral oseltamivir treatment of influenza in children. Pediatr Infect Dis J 20:127–133
5. Gubareva LV (2004) Molecular mechanisms of influenza virus resistance to neuraminidase inhibitors. Virus Res 103:199–203
6. Weinstock DM, Gubareva LV, Zuccotti G (2003) Prolonged shedding of multidrug-resistant influenza A virus in an immunocompromised patient. N Engl J Med 348:867–868
7. Sheu TG, Deyde VM, Okomo-Adhiambo M, Garten RJ, Xu X, Bright RA, Butler EN, Wallis TR, Klimov AI, Gubareva LV (2008) Surveillance for neuraminidase inhibitor resistance among human influenza A and B viruses circulating worldwide from 2004 to 2008. Antimicrob Agents Chemother 52:3284–3292
8. Stephenson I, Democratis J, Lackenby A, McNally T, Smith J, Pareek M, Ellis J, Bermingham A, Nicholson K, Zambon M (2009) Neuraminidase Inhibitor Resistance after Oseltamivir Treatment of Acute Influenza A and B in Children. Clin Infect Dis 48:389–396
9. Le QM, Kiso M, Someya K, Sakai YT, Nguyen TH, Nguyen KH, Pham ND, Ngyen HH, Yamada S, Muramoto Y et al (2005) Avian flu: isolation of drug-resistant H5N1 virus. Nature 437:1108
10. de Jong MD, Tran TT, Truong HK, Vo MH, Smith GJ, Nguyen VC, Bach VC, Phan TQ, Do QH, Guan Y et al (2005) Oseltamivir resistance during treatment of influenza A (H5N1) infection. N Engl J Med 353:2667–2672
11. Normile D, Enserink M (2007) Avian influenza. With change in the seasons, bird flu returns. Science 315:448
12. Kiso M, Mitamura K, Sakai-Tagawa Y, Shiraishi K, Kawakami C, Kimura K, Hayden FG, Sugaya N, Kawaoka Y (2004) Resistant influenza A viruses in children treated with oseltamivir: descriptive study. Lancet 364:759–765
13. Covington E, Mendel DB, Escarpe P, Tai CY, Soder-barg K, Roberts N (2000) Phenotypic and genotypic assay of influenza virus neuraminidase indicates a low incidence of viral drug resistance during treatment with oseltamivir. J Clin Virol 18:251–269 (P-326)
14. Ison MG, Gubareva LV, Atmar RL, Treanor J, Hayden FG (2006) Recovery of drug-resistant influenza virus from immunocompromised patients: a case series. J Infect Dis 193:760–764
15. Baz M, Abed Y, McDonald J, Boivin G (2006) Characterization of multidrug-resistant influenza A/H3N2 viruses shed during 1 year by an immunocompromised child. Clin Infect Dis 43:1555–1561
16. Abed Y, Baz M, Boivin G (2009) A novel neuraminidase deletion mutation conferring resistance to oseltamivir in clinical influenza A/H3N2 virus. J Infect Dis 199:180–183
17. Ives J, Carr J, Roberts NA, Tai CY, Mendel DB, Kelly L, Lambkin R, Oxford J (2000) An oseltamivir treatment-selected influenza A/N2 virus with a R292K mutation in the neuraminidase gene has reduced infectivity in vivo. J Clin Virol 18:251–269 (P-321)
18. Gubareva LV, Matrosovich MN, Brenner MK, Bethell RC, Webster RG (1998) Evidence for zanamivir resistance in an immunocompromised child infected with influenza B virus. J Infect Dis 178:1257–1262
19. Hatakeyama S, Sugaya N, Ito M, Yamazaki M, Ichikawa M, Kimura K, Kiso M, Shimizu H, Kawakami C, Koike K et al (2007) Emergence of influenza B viruses with reduced sensitivity to neuraminidase inhibitors. JAMA 297:1435–1442

20. Ison MG, Gnann JW Jr, Nagy-Agren S, Treannor J, Paya C, Steigbigel R, Elliott M, Weiss HL, Hayden FG (2003) Safety and efficacy of nebulized zanamivir in hospitalized patients with serious influenza. Antivir Ther 8:183–190
21. Ilyushina NA, Govorkova EA, Gray TE, Bovin NV, Webster RG (2008) Human-like receptor specificity does not affect the neuraminidase inhibitor susceptibility of H5N1 influenza viruses. PLoS Pathog 4:e1000043
22. Wang MZ, Tai CY, Mendel DB (2002) Mechanism by which mutations at his274 alter sensitivity of influenza a virus n1 neuraminidase to oseltamivir carboxylate and zanamivir. Antimicrob Agents Chemother 46:3809–3816
23. Abed Y, Goyette N, Boivin G (2004) A reverse genetics study of resistance to neuraminidase inhibitors in an influenza A/H1N1 virus. Antivir Ther 9:577–581
24. Yen HL, Ilyushina NA, Salomon R, Hoffmann E, Webster RG, Govorkova EA (2007) Neuraminidase inhibitor-resistant recombinant A/Vietnam/1203/04 (H5N1) influenza viruses retain their replication efficiency and pathogenicity *in vitro* and *in vivo*. J Virol 81:12418–12426
25. Abed Y, Nehme B, Baz M, Boivin G (2008) Activity of the neuraminidase inhibitor A-315675 against oseltamivir-resistant influenza neuraminidases of N1 and N2 subtypes. Antiviral Res 77:163–166
26. Zurcher T, Yates PJ, Daly J, Sahasrabudhe A, Walters M, Dash L, Tisdale M, McKimm-Breschkin JL (2006) Mutations conferring zanamivir resistance in human influenza virus N2 neuraminidases compromise virus fitness and are not stably maintained *in vitro*. J Antimicrob Chemother 58:723–732
27. Jackson D, Barclay W, Zurcher T (2005) Characterization of recombinant influenza B viruses with key neuraminidase inhibitor resistance mutations. J Antimicrob Chemother 55:162–169
28. Yen HL, Herlocher LM, Hoffmann E, Matrosovich MN, Monto AS, Webster RG, Govorkova EA (2005) Neuraminidase inhibitor-resistant influenza viruses may differ substantially in fitness and transmissibility. Antimicrob Agents Chemother 49:4075–4084
29. Bouvier NM, Lowen AC, Palese P (2008) Oseltamivir-resistant influenza A viruses are transmitted efficiently among guinea pigs by direct contact but not by aerosol. J Virol 82: 10052–10058
30. Gubareva LV, Bethell R, Hart GJ, Murti KG, Penn CR, Webster RG (1996) Characterization of mutants of influenza A virus selected with the neuraminidase inhibitor 4-guanidino-Neu5Ac2en. J Virol 70:1818–1827
31. Gubareva LV, Robinson MJ, Bethell RC, Webster RG (1997) Catalytic and framework mutations in the neuraminidase active site of influenza viruses that are resistant to 4-guanidino-Neu5Ac2en. J Virol 71:3385–3390
32. Blick TJ, Tiong T, Sahasrabudhe A, Varghese JN, Colman PM, Hart GJ, Bethell RC, McKimm-Breschkin JL (1995) Generation and characterization of an influenza virus neuraminidase variant with decreased sensitivity to the neuraminidase-specific inhibitor 4-guanidino-Neu5Ac2en. Virology 214:475–484
33. Staschke KA, Colacino JM, Baxter AJ, Air GM, Bansal A, Hornback WJ, Munroe JE, Laver WG (1995) Molecular basis for the resistance of influenza viruses to 4-guanidino-Neu5Ac2en. Virology 214:642–646
34. Barnett JM, Cadman A, Burrell FM, Madar SH, Lewis AP, Tisdale M, Bethell R (1999) In vitro selection and characterisation of influenza B/Beijing/1/87 isolates with altered susceptibility to zanamivir. Virology 265:286–295
35. Tai CY, Escarpe PA, Sidwell RW, Williams MA, Lew W, Wu H, Kim CU, Mendel DB (1998) Characterization of human influenza virus variants selected in vitro in the presence of the neuraminidase inhibitor GS 4071. Antimicrob Agents Chemother 42:3234–3241
36. Yen HL, Hoffmann E, Taylor G, Scholtissek C, Monto AS, Webster RG, Govorkova EA (2006) Importance of neuraminidase active-site residues to the neuraminidase inhibitor resistance of influenza viruses. J Virol 80:8787–8795
37. Mishin VP, Hayden FG, Gubareva LV (2005) Susceptibilities of antiviral-resistant influenza viruses to novel neuraminidase inhibitors. Antimicrob Agents Chemother 49:4515–4520

38. Russell RJ, Haire LF, Stevens DJ, Collins PJ, Lin YP, Blackburn GM, Hay AJ, Gamblin SJ, Skehel JJ (2006) The structure of H5N1 avian influenza neuraminidase suggests new opportunities for drug design. Nature 443:45–49
39. Varghese JN, Smith PW, Sollis SL, Blick TJ, Sahasrabudhe A, McKimm-Breschkin JL, Colman PM (1998) Drug design against a shifting target: a structural basis for resistance to inhibitors in a variant of influenza virus neuraminidase. Structure 6:735–746
40. Collins PJ, Haire LF, Lin YP, Liu J, Russell RJ, Walker PA, Skehel JJ, Martin SR, Hay AJ, Gamblin SJ (2008) Crystal structures of oseltamivir-resistant influenza virus neuraminidase mutants. Nature 453:1258–1261
41. Carr J, Ives J, Kelly L, Lambkin R, Oxford J, Mendel D, Tai L, Roberts N (2002) Influenza virus carrying neuraminidase with reduced sensitivity to oseltamivir carboxylate has altered properties in vitro and is compromised for infectivity and replicative ability in vivo. Antiviral Res 54:79–88
42. McKimm-Breschkin JL, Sahasrabudhe A, Blick TJ, McDonald M, Colman PM, Hart GJ, Bethell RC, Varghese JN (1998) Mutations in a conserved residue in the influenza virus neuraminidase active site decreases sensitivity to Neu5Ac2en-derived inhibitors. J Virol 72:2456–2462
43. Rameix-Welti MA, Agou F, Buchy P, Mardy S, Aubin JT, Veron M, van der Werf S, Naffakh N (2006) Natural variation can significantly alter sensitivity to oseltamivir of Influenza A(H5N1) viruses. Antimicrob Agents Chemother 50:3809–3815
44. McKimm-Breschkin JL, McDonald M, Blick TJ, Colman PM (1996) Mutation in the influenza virus neuraminidase gene resulting in decreased sensitivity to the neuraminidase inhibitor 4-guanidino-Neu5Ac2en leads to instability of the enzyme. Virology 225:240–242
45. Baz M, Abed Y, Boivin G (2007) Characterization of drug-resistant recombinant influenza A/H1N1 viruses selected in vitro with peramivir and zanamivir. Antiviral Res 74:159–162
46. Ives JA, Carr JA, Mendel DB, Tai CY, Lambkin R, Kelly L, Oxford JS, Hayden FG, Roberts NA (2002) The H274Y mutation in the influenza A/H1N1 neuraminidase active site following oseltamivir phosphate treatment leave virus severely compromised both *in vitro* and *in vivo*. Antiviral Res 55:307–317
47. Ives J, Carr J, Roberts NA, Tai CY, Mendel DB, Kelly L, Lambkin R, Oxford J (2000) An oseltamivir treatment-selected influenza A/Wuhan/359/95 virus with a E119V mutation in the neuraminidase gene has reduced infectivity in vivo. J Clin Virol 18:251–269 (P-330)
48. Abed Y, Baz M, Boivin G (2006) Impact of neuraminidase mutations conferring influenza resistance to neuraminidase inhibitors in the N1 and N2 genetic backgrounds. Antivir Ther 11:971–976
49. Herlocher ML, Truscon R, Elias S, Yen HL, Roberts NA, Ohmit SE, Monto AS (2004) Influenza viruses resistant to the antiviral drug oseltamivir: transmission studies in ferrets. J Infect Dis 190:1627–1630
50. Herlocher ML, Carr J, Ives J, Elias S, Truscon R, Roberts N, Monto AS (2002) Influenza virus carrying an R292K mutation in the neuraminidase gene is not transmitted in ferrets. Antiviral Res 54:99–111
51. Rameix-Welti MA, Enouf V, Cuvelier F, Jeannin P, van der Werf S (2008) Enzymatic properties of the neuraminidase of seasonal H1N1 influenza viruses provide insights for the emergence of natural resistance to oseltamivir. PLoS Pathog 4:e1000103
52. Potier M, Mameli L, Belisle M, Dallaire L, Melancon SB (1979) Fluorometric assay of neuraminidase with a sodium (4-methylumbelliferyl-alpha-D-N-acetylneuraminate) substrate. Anal Biochem 94:287–296
53. Gubareva LV, Webster RG, Hayden FG (2002) Detection of influenza virus resistance to neuraminidase inhibitors by an enzyme inhibition assay. Antiviral Res 53:47–61
54. Buxton RC, Edwards B, Juo RR, Voyta JC, Tisdale M, Bethell RC (2000) Development of a sensitive chemiluminescent neuraminidase assay for the determination of influenza virus susceptibility to zanamivir. Anal Biochem 280:291–300

55. Wetherall NT, Trivedi T, Zeller J, Hodges-Savola C, McKimm-Breschkin JL, Zambon M, Hayden FG (2003) Evaluation of neuraminidase enzyme assays using different substrates to measure susceptibility of influenza virus clinical isolates to neuraminidase inhibitors: report of the neuraminidase inhibitor susceptibility network. J Clin Microbiol 41:742–750
56. Kati WM, Montgomery D, Carrick R, Gubareva L, Maring C, McDaniel K, Steffy K, Molla A, Hayden F, Kempf D et al (2002) In vitro characterization of A-315675, a highly potent inhibitor of A and B strain influenza virus neuraminidases and influenza virus replication. Antimicrob Agents Chemother 46:1014–1021
57. McSharry JJ, McDonough AC, Olson BA, Drusano GL (2004) Phenotypic drug susceptibility assay for influenza virus neuraminidase inhibitors. Clin Diagn Lab Immunol 11:21–28
58. Tisdale M (2000) Monitoring of viral susceptibility: new challenges with the development of influenza NA inhibitors. Rev Med Virol 10:45–55
59. Zambon M, Hayden FG (2001) Position statement: global neuraminidase inhibitor susceptibility network. Antiviral Res 49:147–156
60. Hatakeyama S, Sakai-Tagawa Y, Kiso M, Goto H, Kawakami C, Mitamura K, Sugaya N, Suzuki Y, Kawaoka Y (2005) Enhanced expression of an alpha2,6-linked sialic acid on MDCK cells improves isolation of human influenza viruses and evaluation of their sensitivity to a neuraminidase inhibitor. J Clin Microbiol 43:4139–4146
61. Matrosovich M, Matrosovich T, Carr J, Roberts NA, Klenk HD (2003) Overexpression of the alpha-2,6-sialyltransferase in MDCK cells increases influenza virus sensitivity to neuraminidase inhibitors. J Virol 77:8418–8425
62. Matrosovich MN, Matrosovich TY, Gray T, Roberts NA, Klenk HD (2004) Human and avian influenza viruses target different cell types in cultures of human airway epithelium. Proc Natl Acad Sci USA 101:4620–4624
63. Carr MJ, Sayre N, Duffy M, Connell J, Hall WW (2008) Rapid molecular detection of the H275Y oseltamivir resistance gene mutation in circulating influenza A (H1N1) viruses. J Virol Methods 153:257–262
64. Chutinimitkul S, Suwannakarn K, Chieochansin T, le Mai Q, Damrongwatanapokin S, Chaisingh A, Amonsin A, Landt O, Songserm T, Theamboonlers A et al (2007) H5N1 Oseltamivir-resistance detection by real-time PCR using two high sensitivity labeled TaqMan probes. J Virol Methods 139:44–49
65. Deyde VM, Nguyen T, Bright RA, Balish A, Shu B, Lindstrom S, Klimov AI, Gubareva LV (2009) Detection of molecular markers of antiviral resistance in influenza A (H5N1) viruses using a pyrosequencing method. Antimicrob Agents Chemother 53:1039–1047
66. Deyde VM, Okomo-Adhiambo M, Sheu TG, Wallis TR, Fry A, Dharan N, Klimov AI, Gubareva LV (2009) Pyrosequencing as a tool to detect molecular markers of resistance to neuraminidase inhibitors in seasonal influenza A viruses. Antiviral Res 81:16–24
67. Duwe S, Schweiger B (2008) A new and rapid genotypic assay for the detection of neuraminidase inhibitor resistant influenza A viruses of subtype H1N1, H3N2, and H5N1. J Virol Methods 153:134–141
68. Lackenby A, Democratis J, Siqueira MM, Zambon MC (2008) Rapid quantitation of neuraminidase inhibitor drug resistance in influenza virus quasispecies. Antivir Ther 13:809–820
69. Bright RA, Medina MJ, Xu X, Perez-Oronoz G, Wallis TR, Davis XM, Povinelli L, Cox NJ, Klimov AI (2005) Incidence of adamantane resistance among influenza A (H3N2) viruses isolated worldwide from 1994 to 2005: a cause for concern. Lancet 366:1175–1181
70. Aoki FY, Boivin G, Roberts N (2007) Influenza virus susceptibility and resistance to oseltamivir. Antivir Ther 12:603–616
71. Oo C, Hill G, Dorr A, Liu B, Boellner S, Ward P (2003) Pharmacokinetics of anti-influenza prodrug oseltamivir in children aged 1–5 years. Eur J Clin Pharmacol 59:411–415
72. Ward P, Small I, Smith J, Suter P, Dutkowski R (2005) Oseltamivir (Tamiflu) and its potential for use in the event of an influenza pandemic. J Antimicrob Chemother 55(Suppl 1):i5–i21

73. McKimm-Breschkin J, Trivedi T, Hampson A, Hay A, Klimov A, Tashiro M, Hayden F, Zambon M (2003) Neuraminidase sequence analysis and susceptibilities of influenza virus clinical isolates to zanamivir and oseltamivir. Antimicrob Agents Chemother 47:2264–2272
74. Boivin G, Goyette N (2002) Susceptibility of recent Canadian influenza A and B virus isolates to different neuraminidase inhibitors. Antiviral Res 54:143–147
75. Hurt AC, Barr IG, Hartel G, Hampson AW (2004) Susceptibility of human influenza viruses from Australasia and South East Asia to the neuraminidase inhibitors zanamivir and oseltamivir. Antiviral Res 62:37–45
76. Mungall BA, Xu X, Klimov A (2004) Surveillance of influenza isolates for susceptibility to neuraminidase inhibitors during the 2000–2002 influenza seasons. Virus Res 103:195–197
77. Monto AS, McKimm-Breschkin JL, Macken C, Hampson AW, Hay A, Klimov A, Tashiro M, Webster RG, Aymard M, Hayden FG et al (2006) Detection of influenza viruses resistant to neuraminidase inhibitors in global surveillance during the first 3 years of their use. Antimicrob Agents Chemother 50:2395–2402
78. Ferraris O, Kessler N, Lina B (2005) Sensitivity of influenza viruses to zanamivir and oseltamivir: a study performed on viruses circulating in France prior to the introduction of neuraminidase inhibitors in clinical practice. Antiviral Res 68:43–48
79. Ferraris O, Kessler N, Valette M, Lina B (2006) Evolution of the susceptibility to antiviral drugs of A/H3N2 influenza viruses isolated in France from 2002 to 2005. Vaccine 24:6656–6659
80. Escuret V, Frobert E, Bouscambert-Duchamp M, Sabatier M, Grog I, Valette M, Lina B, Morfin F, Ferraris O (2008) Detection of human influenza A (H1N1) and B strains with reduced sensitivity to neuraminidase inhibitors. J Clin Virol 41:25–28
81. Hurt AC, Barr IG (2008) Influenza viruses with reduced sensitivity to the neuraminidase inhibitor drugs in untreated young children. Commun Dis Intell 32:57–62
82. van der Vries E, van den Berg B, Schutten M (2008) Fatal oseltamivir-resistant influenza virus infection. N Engl J Med 359:1074–1076
83. Nichols WG, Guthrie KA, Corey L, Boeckh M (2004) Influenza infections after hematopoietic stem cell transplantation: risk factors, mortality, and the effect of antiviral therapy. Clin Infect Dis 39:1300–1306
84. de Jong MD, Simmons CP, Thanh TT, Hien VM, Smith GJ, Chau TN, Hoang DM, Chau NV, Khanh TH, Dong VC et al (2006) Fatal outcome of human influenza A (H5N1) is associated with high viral load and hypercytokinemia. Nat Med 12:1203–1207
85. Monto AS, Comanor L, Shay DK, Thompson WW (2006) Epidemiology of pandemic influenza: use of surveillance and modeling for pandemic preparedness. J Infect Dis 194(Suppl 2): S92–97
86. Taylor NR, von Itzstein M (1996) A structural and energetics analysis of the binding of a series of N-acetylneuraminic-acid-based inhibitors to influenza virus sialidase. J Comput Aided Mol Des 10:233–246
87. Hauge SH, Dudman S, Borgen K, Lackenby A, Hungnes O (2009) Oseltamivir-resistant influenza viruses A (H1N1), Norway, 2007–08. Emerg Infect Dis 15:155–162
88. Tamura D, Mitamura K, Yamazaki M, Fujino M, Nirasawa M, Kimura K, Kiso M, Shimizu H, Kawakami C, Hiroi S et al (2009) Oseltamivir-resistant influenza A viruses circulating in Japan. J Clin Microbiol 47:1424–1427
89. Dharan NJ, Gubareva LV, Meyer JJ, Okomo-Adhiambo M, McClinton RC, Marshall SA, St George K, Epperson S, Brammer L, Klimov AI et al (2009) Infections with oseltamivir-resistant influenza A(H1N1) virus in the United States. JAMA 301:1034–1041
90. Vilchez RA, McCurry K, Dauber J, Lacono A, Griffith B, Fung J, Kusne S (2002) Influenza virus infection in adult solid organ transplant recipients. Am J Transplant 2:287–291
91. Kim YJ, Boeckh M, Englund JA (2007) Community respiratory virus infections in immunocompromised patients: hematopoietic stem cell and solid organ transplant recipients, and individuals with human immunodeficiency virus infection. Semin Respir Crit Care Med 28:222–242

Index

M. von Itzstein (ed.), *Influenza Virus Sialidase - A Drug Discovery Target*,
Milestones in Drug Therapy, DOI 10.1007/978-3-7643-8927-7,

MIX
Papier aus verantwortungsvollen Quellen
Paper from responsible sources
FSC® C105338

If you have any concerns about our products,
you can contact us on
ProductSafety@springernature.com

In case Publisher is established outside the EU,
the EU authorized representative is:
Springer Nature Customer Service Center GmbH
Europaplatz 3, 69115 Heidelberg, Germany

Printed by Libri Plureos GmbH
in Hamburg, Germany